SPACEX STARSHIP:

ELON MUSK'S

ROCKET TO MARS

By Jon Binder

Legal Notice:

This book is copyright protected. It is only for personal use. You cannot amend, distribute, sell, use, quote or paraphrase any part, or the content within this book, without the consent of the author or publisher.

Disclaimer Notice:

Please note the information contained within this document is for educational and entertainment purposes only.

All effort has been executed to present accurate, up to date, reliable, complete information. No warranties of any kind are declared or implied. Readers acknowledge that the author is not engaged in the rendering of legal, financial, medical or professional advice.

The content within this book has been derived from various sources. Please consult a licensed professional before attempting any techniques outlined in this book.

By reading this document, the reader agrees that under no circumstances is the author responsible for any losses, direct or indirect, that are incurred as a result of the use of the information contained within this document, including, but not limited to, errors, omissions, or inaccuracies.

Table of Contents

Introduction

Space is cold, dark and unforgiving. Since there is no air in space, if you scream, it makes no sound.

In the coming years, SpaceX will launch a crew of 24 people through space and onto Mars, in a new kind of rocket called Starship.

In the year 2038, the space exploration company will divide the crew into two groups of 12 and send them onto Mars Base Alpha, the first human outpost on another planet. At 80 million miles away, Mars is both desolate, and at the same time full of limitless potential. Red and barron, Mars holds limitless possibilities.

These two Starships, along with several cargo-carrying companion rockets will begin the months-long journey to make humans a multiplanetary species.

Since the ideal alignment for Earth and Mars only happens once every two years, there is only a small window to get everything right. Timing is everything.

The future of space exploration is unlike anything we've ever seen before. Although we've been exploring space since the 1950s, today, for the first time ever, we are going to be using supersized spacecraft to supercharge our journey into the solar system.

As far as we know, Earth is the only planet in the known universe to contain any form of life. However, these 24 bold adventurers will become the first of what will eventually become a Mars colony of tens of thousands of people willing to leave everything behind to forge a new life on a new planet.

But before going to Mars, Starship will first complete several lunar missions to establish the Artemis Base Camp on the moon. Artemis will allow astronauts to stay for longer periods of time and complete important research projects to prepare for people living on Mars.

The Starship vehicle that's being developed by SpaceX will be able to launch more payload, have more power, and fly more often than any rocket ever produced. The last Saturn 5 rocket was launched by NASA in 1973, and since then, there has never been a more powerful rocket, until now. Starship can be refueled to be launched over and over again. It will become the workhorse used to carry thousands of people and millions of tons of cargo throughout the solar system.

These are not small advances in rocketry or spacecraft. These are truly transformational technologies that are giving us completely new capabilities and changing the paradigm for space exploration.

In 2021, NASA officially selected SpaceX to build the first human lunar lander to venture to the moon. This greatly angered SpaceX CEO Elon Musk's biggest competitor, Amazon founder Jeff Bezos. Bezos and his competing company Blue

Origin, have been competing with and hiring talent away from SpaceX for over 20 years. The Amazon founder has been funding his private rocket company personally since September of 2000, selling $1 billion dollars of Amazon stock every year. Blue Origin is determined to beat SpaceX at all costs, but so far hasn't been able to catch up to Musk and his innovative rockets.

Starship has enabled us to do something that for the first time in the history of our planet has only been the stuff of myth or science fiction. In over four and a half billion years, we are on the cusp of sending humans to build a home away from earth. But first, we need to send uncrewed Starships to Mars to prove that we can both safely land safely and also build the infrastructure, before we send humans.

We will use the awesome payload capacity of Starship to send the building blocks and essential elements needed to sustain a human presence on Mars. Starting with what is called ISRU or In-Situ Resource Utilization systems. ISTU basically means living off the land. If we want to have a self-sustaining presence on Mars, we cannot be reliant on supplies from Earth. If the resupply rockets stop coming for any reason, the people living on Mars need to have everything they need in order to survive and thrive.

Our ancestors, for tens of thousands of years, have been learning how to use local resources to do everything from building tools, growing food and generating energy. We live on a pretty nice planet, and so far we have everything that we need in order to live our lives. But Mars is different. Mars is cold and unforgiving. If you run out of food, or water, or oxygen, you're dead. So we have to be very wise about how to survive on Mars.

And one of the main ways we're going to start is by using ice as a resource. We already know that Mars has lots and lots of water ice. There is both ground ice and there's rock-covered glaciers for us to use. Traditionally, when people talk about sending humans to Mars, we're talking about just a few people and maybe even a small rover. But Starship is so big and so transformational that we can now discuss sending not only supplies but also heavy-duty construction machinery to build the infrastructure that we need for a large-scale presence on Mars.

The Starship rocket, and its Super-Heavy booster are a fully reusable transportation system. That means that you don't have to build a new rocket every time you want to fly. This vastly reduces the cost of each flight and lets us fly more often. Historically, rockets have only been able to be used once. But if you think about it, would you ever build an airplane, fly it once and then throw it away? If so, the cost to fly anything would be so expensive and so rare that any air travel would have to be reserved for extremely special occasions.

For example, consider the last rover that was sent to Mars. The launch cost for that mission was $243 million dollars. That is about $100,000 per pound. SpaceX is aiming to have a total launch cost for Starship on the order of a couple million dollars total. That comes out to roughly $900 per pound. By comparison, that's $100,000 for NASA versus $900 for SpaceX. Not only a massive price difference, but also so much more affordable that it opens up a brand new realm of possibilities for space travel.

It has taken the entire history of our planet to reach the point where we're at now. It's an amazing time to be alive, and what we choose to do next will forever change the course of human history.

Now is the time to seize the opportunity to expand human consciousness throughout the solar system. If we act decisively, and with a little bit of luck, we can send humans on the ride of a lifetime. Venturing onto Mars, we will change the course of history forever.

Chapter 1:

A Brief History of SpaceX

"If humanity doesn't land on Mars in my lifetime, I would be very disappointed. "

-Elon Musk

Boca Chica, South Texas: It sits there, gleaming silver in the sunlight like some icon from a science fiction movie. Starship. Its very name describes everything about its function and the dream of its creator—to travel to the stars. The ambition of Elon Musk, founder and CEO of SpaceX has reached new heights with a goal to send Starship to the Moon, Mars, and beyond to further human exploration and create a self-sustaining Martian civilization of over 1 million people.

Little did the public or the space industry anticipate the meteoric rise of SpaceX, an upstart, private, rocket company run by a billionaire with no prior space engineering experience. But Elon Musk intuitively understood what it took to become successful with a space agency, and that was cutting the cost of getting payloads to space. Cost is the main concern of companies who want to launch their satellites into orbit.

But Musk went farther than anyone, not only cutting costs but also advancing innovation through engineering, from scratch, a new type of reusable rocket that would also land itself on a 3-legged tripod. This type of engineering at first seemed strange to traditional space agencies and was considered science fiction, expensive, wasteful and unnecessary.

But they were all proved wrong. The extravagance of the Falcon rockets' reusability and turnaround to the next flight demonstrated that such a rocket system was not only possible, but actually vastly cheaper to run than what national space agencies could achieve. Today, even NASA hires SpaceX for crew and resupply missions to the International Space Station and for launching satellites. They are even considering SpaceX as the number 1 contender to return to the moon with a series of manned moon missions.

SpaceX is now a successful, fully-fledged space agency building its own rockets and engines, launching from its own 'starbases', providing the world with a new communications satellite network called Starlink, sending private and commercial crews into space, and orbital test flights, missions to the Moon, and developing the even-more-ambitious target of putting one million people on Mars with the most powerful rocket ever built, called Starship.

Before you is the story of Starship, starting with the history of SpaceX's beginnings, from its current space programs to the forward-thinking Starship. We go in depth into Starship's technical makeup, the high-altitude testing, and planned orbital tests. Looking ahead, the proposed missions to the Moon and Mars are assessed. It is known Musk is quite keen to fly one million people to Mars, but is this a fulfillable mission? This leads to examinations of the criticism of both SpaceX and issues with Starship. How will one million people travel to Mars? How will they live and work on the Red Planet? It is a very bold, humanity-expanding goal, and one Elon Musk and SpaceX are determined to accomplish. The prize for winning company contracts with NASA or other commercial agencies is enormous and so are the benefits of being first to the Moon, Mars, or beyond. So, we can look forward to the future of SpaceX and Starship.

Starship is the tip of the spear of humanity's long-term desire to explore and live as a multiplanetary species. It will herald in the next evolutionary step forward for our civilization. Today, Starship outranks any other space transportation system to achieve this goal within our lifetime. This silver dream upon which humanity will light up the skies as we thrust our way into the cosmos will be a beacon of hope and survival. And if any doubt remains about our future path, no two words have come together to better assert the drive, challenge, and ambition of humanity in space than "Starship".

So, as science fiction comes to life when Starship rockets aloft into the heavens, the small mouth of the Rio Grande, Texas, will let loose the great roar heard around the world. Go, Starship, go!

Space Exploration Technologies was born in 2002, after Elon Musk, frustrated at his attempts to buy a rocket for his various experiments and commercial endeavors, decided to set up his own commercial space agency. And it started with some rodents.

Space Mice

In 2001, Elon Musk sold his internet payments company PayPal, leaving him with lots of time and money on his hands for a new venture. But instead of starting another internet company, Musk had his sights set on something bigger. He moved his family to Los Angeles in order to reinvigorate his life after a near-death experience with malaria, contracted while in South Africa, and to renew his interest in space. Southern California is a hub of aerospace engineering and the perfect place to start a new venture focused on space. It was here he first encountered the Mars Society, a nonprofit organization dedicated to colonizing Mars. After donating generously at their 2001 fundraiser and listening to interesting space and engineering ideas, Musk joined their board of directors.

The Mars Society had an idea for an experiment called the "Translife Mission", which would see mice being sent into orbit in a spinning capsule set at one-third of Earth's gravity to simulate Mars', so the mice could live comfortably and also reproduce. Musk's idea was to actually send the mice to Mars. Even though he was ribbed for the idea, this was the beginning of Musk's burning ambition to set humans on an interplanetary mission of exploration and colonization.

Musk was disappointed NASA had no plans for a Mars mission. After resigning from the Mars Society, Musk set up his own enterprise, Life to Mars Foundation. His meetings with scientists, engineers, and academics bore fruit with his "mice to Mars" idea transforming into the Mars Oasis project. Musk's plan was to purchase a rocket for $20 to $30 million and use it to launch a small, robotic, experimental greenhouse to Mars. The plants within a chamber in the rocket would interact with Martian soil with the intent to produce oxygen, and the effects would be studied. Videos of the experiment would be beamed back to Earth to inspire kids and adults to grow their own plants and be inspired by space travel. The main problem was that space experts saw the mission costing at least $200 million.

But undeterred, this sent Musk searching for rockets in Russia for overhauled, intercontinental ballistic missiles (ICBMs), which ultimately fell through; and to Paris and London, meeting smaller astronautic companies. It was only in 2002 when Musk told his small team they could build the rockets themselves. During the past year, Musk had not been idle in just spending money on futile space endeavors; rather, he had been studying the space industry, its costs, and physics, and saw he could build exactly what he required and for much cheaper than current rockets. And most importantly, he had been inspired to strive beyond mice in space. It was time for humanity to return to space exploration.

SpaceX is Born

So Musk, Jim Cantrell, Tom Mueller, and aerospace engineer Chris Thompson became the first of many of SpaceX to set up in a large warehouse in El Segundo, Los Angeles. As SpaceX acquired more engineers and employees, its reputation grew, and the buzz about a new, private space agency caught the attention of many.

From the offset, SpaceX's mission has been to make humanity multiplanetary. SpaceX has striven to accelerate the innovations in the space industry, building on the Falcon 9s, Falcon Heavy, and the Dragons—each one developing out of the other and culminating in Starship. The fully-reusable Falcon rockets, capable of landing back on Earth, are a step forward which space experts envision as a breakthrough for Moon and Mars missions. The reusability of the rockets greatly reduces the cost of space travel, eliminating rocket-building costs and time between each launch.

In 2012, the Dragon spacecraft became the first private craft to deliver cargo to and from the International Space Station (ISS), with crewed missions following in 2020, another first for a commercial company.

Musk's motivation was to have SpaceX become the "Southwest Airlines of Space". By this, he meant having rockets that could fly multiple missions over their lifetimes, much like a commercial airliner.

The building costs of a Falcon rocket and an airline was about the same, but the airliner could fly far more times. Musk wanted SpaceX to emulate an airline company, building its own engines and contracting out for other specific parts and off-the-shelf parts, creating reusable rockets with quick turnaround times between launches to greatly reduce the cost of traveling to space by a hundredfold.

SpaceX is not situated in one state. It occupies three states which are synonymous with American spaceflight, namely: California, Texas, and Florida. Three core stages of SpaceX operations are the building, testing, and launching.

California has both build and launch facilities. SpaceX's headquarters are in Hawthorne, California, where the engineers design and build its spacecraft. SpaceX is called a "vertically integrated company", meaning the entire process of building their spacecraft is located under one roof, a distinct advantage where designers, engineers, and builders are clustered together for quick iterations on ideas and builds—again, saving on time and costs of operational matters.

McGregor, Texas hosts SpaceX's engines, vehicle structures, and systems testing operations at its 4,000-acre, high-tech rocket development facility. The 16 purpose-built test stands certify every Merlin engine and Draco thruster that power the Falcon 9 and the Dragon spacecraft, respectively.

In regards to flight testing and contracted launches, as of 2020, SpaceX enjoys the privilege of being able to lease or use four launch centers:

1. Cape Canaveral Space Launch Complex 40 (SLC-40) for United States Space Force national security launches. Cape Canaveral's position on the southeast coast of the US also allows for launches of communication satellites in low, medium, and geostationary orbits and supply missions to the ISS and departure launches to the Moon and beyond.

2. Vandenberg Space Force Base Space Launch Complex 4E (SLC-4E) for polar launches. This California base allows for

launches into high inclination and polar orbits for communication, defense, and Earth-observing satellites, plus certain Moon missions.

3. Kennedy Space Center Launch Complex 39A (LC-39A) for NASA launches.

4. Brownsville South Texas Launch Site (founded 2014)—aka "Starbase"—for commercial and orbital missions. This is Starship's build and development home.

With these launch sites, SpaceX states they can "optimize their launch operations, and reduce launch costs, by dividing their launch missions amongst these four launch facilities."

In January 2020, SpaceX bought two drilling-rig platforms from bankruptcy-hit Valaris plc for $3.5 million each in order to repurpose them as offshore spaceports. This gave the Falcon rockets the flexibility to land on one of SpaceX's autonomous drone ship—or offshore spaceports—out on the ocean or one of the landing pads at Starbase.

After the successful second test flight demonstration of the Falcon 9/Dragon on May 30, 2020 delivered the first astronauts to the International Space Station (ISS) in almost a decade, on November 10, 2020, NASA certified SpaceX's Falcon 9 and Crew Dragon human spaceflight system for flight to and from the ISS. Historically, this was the first ever commercial organization to achieve such a feat. It was also the only space system (spacecraft, rocket, and ground support systems) since the space shuttle 40 years prior to be certified by NASA.

On November 14, 2020, SpaceX launched their first crew rotation mission (Crew-1) to the ISS as part of NASA's Commercial Crew Program. NASA permitted SpaceX to lease and launch from the historic Launch Complex 39A at Kennedy Space Center in Florida, which had previously launched both Apollo-era Saturn V rockets and the Space Shuttle. If they had not done so already, SpaceX has cemented their legacy, being the only private company to operate commercially from NASA facilities.

SpaceX has gone from a novelty, outsider, space cargo launcher to the premier, human spaceflight agency with a "core mission" to continue humanity's journey to the stars.

Starlink

Starlink is SpaceX's system of what will be over 4,000 satellites orbiting Earth, enabling access to the Internet in normally inaccessible places with no connection. On October 24, 2020, SpaceX achieved 100 successful spaceflights with its Starlink missions, highlighting the Falcon 9's reliability and reusability.

In another notable first, on September 25, 2020, the United States Space Force's Space and Missile Systems Center (SMC) contracted SpaceX to launch previously-flown Falcon 9 boosters on GPS missions and to recover the first-stage booster. Additionally, Space Force also certified and chose SpaceX to perform vital National Security Space Launch (NSSL) missions for five years. With its fleet of Falcon 9 and Falcon Heavy, SpaceX had positioned itself as the go-to space agency for private, commercial, and military spaceflight missions. To enable and secure contracts with NASA and Space Force, SpaceX had invested their own funds into upgrading their spacecraft fleet and related systems and infrastructure, which not only streamlined SpaceX's operations, but also with NASA and the United States Air Force. The Space Force using SpaceX's reusable space fleet has saved the American taxpayer billions of dollars.

SpaceX has opened the door for the next generation of private space agencies to enter the 21st Century Space Race, beyond Blue Origin, Boeing, United Launch Alliance, Sierra Nevada Corporation, and Orbital. Just like the cargo ships of times past and airliners of today, space travel will become a more common mode of travel and create innovative travel options across Earth. SpaceX, like its terrestrial forebears, will continue to develop, making spaceflight the new frontier. And the next phase of their interplanetary mission is Starship.

Chapter 2:

Starship is Born

"Mars is key to humanity's future in space. It is the closest planet that has all the resources needed to support life and technological civilization. Its complexity uniquely demands the skills of human explorers, who will pave the way for human settlers."

-Robert Zubrin

Starship will be able to launch from the newly-constructed Starbases in Texas and at the Kennedy Space Center in Florida. Before taking humans to Mars, Starship will be flight tested by launching satellites and space probes before potentially catering to space tourists and flights to the Moon via NASA's Artemis program. Such missions will complete the shakedown of Starship and also serve to quantify and qualify SpaceX's mission to deliver space flights at a reduced launch cost.

Starship will be the most powerful rocket ever built. Like its predecessors in the Falcon series, Starship will be fully reusable as an orbital rocket, achieving a reduced launch cost and lengthy maintenance between flights. With the ambition to launch more than 100 metric tons (220,000 lb) into low-Earth orbit, which will also carry both crew and cargo to the Moon, Mars, and beyond. This would categorize Starship as a super heavy-lift launch vehicle. Starship has the option to be stacked upon a Super Heavy as the first stage, a booster as the second stage, or act as its own spacecraft. However, stacked together, Starship and Super Heavy would be 120 m (390 ft) high, nine meters (30 ft) taller than the Apollo-era Saturn V. A fully-fueled Starship would amass to about 5,000 tons (11,000,000 lb) and power the Raptor and Raptor Vacuum engines. With its reusability and the capability of SpaceX to quickly manufacture rockets, there could be up to three daily Starship launches. This has led to environmental concerns, which are covered in Chapter 9.

Starship's History

During Falcon's heyday, SpaceX conceived of a fully-reusable, super heavy-lift launch system in 2005. The design and name went through many iterations, such as "Mars Oasis", 'BFR', and the "Interplanetary Transport System", before 'Starship' was initiated.

BFR

Elon Musk first addressed the notion of a large rocket in November 2005, known at the time as 'BFR' ("Big Falcon Rocket" or "Big F**cking Rocket"). It was the first inkling of the Starship concept and powered a larger version of the Merlin engine, known as "Merlin 2", capable of launching 100 tons

(220,000 lb) into low-Earth orbit. However, the BFR was not a reusable space vehicle.

Mars Colonial Transporter

Between 2011 and 2017, expanding on its Mars mission mandate, SpaceX revealed a low-cost Mars mission carrying a modified version of the Dragon 1 capsule called "Red Dragon". However, the Dragon capsule's propulsive landing equipment was judged to be unsafe, and so the BFR project was reviewed and realized as the "Mars Colonial Transporter" (MCT) concept. The MCT would have had thrust from methane-fueled Raptor engines, enabling a carrying capacity of 100 people or 100 tons (220,000 lb) of cargo to Mars.

Interplanetary Transport System

At the 67th International Astronautical Congress, September 27, 2016, Musk announced that SpaceX was developing a new, two-stage, reusable rocket called the "Interplanetary Transport System" (ITS). Only the day before had SpaceX fired-tested the Raptor engine for the first time.

The projected launch capacity was a massive 300 tons (660,000 lb) into low-Earth orbit. The booster stage would be powered by 42 Raptor engines with the spacecraft equipped with nine Raptors. The ITS also introduced the potential of using the spacecraft as a tanker, able to ferry extra fuel into orbit to transfer to another spacecraft. The rocket's fuel tanks would be constructed from carbon composite, storing the Raptor's liquid methane and liquid oxygen.

Back to BFR

Despite the ITS concept seemingly an ambition too far for its many skeptics, SpaceX pressed on. At the 68th Annual International Astronautical Congress in September 2017, Musk announced a revision to the Interplanetary Transport System, reviving the BFR name for the project. BFR was now a reusable launch system, but with a capacity decreased to 150 tons (330,000 lb). The spacecraft's body size was reduced, and the 42 ITS booster engines were whittled down to 31. Though BFR's main purpose was a mission to Mars, it was also to be a versatile workhorse with capabilities to launch satellites into orbit, travel to the ISS, land on the Moon, and fly between Earth-bound spaceports. In a show of commitment to SpaceX, in April 2018, the Mayor of Los Angeles announced plans for a BFR rocket-production facility at the Port of Los Angeles, but ultimately, the plan was shelved in May 2020. However, the seriousness with which SpaceX had dedicated its future to Starship was only growing.

Starship

In September 2018, Japanese billionaire Yusaku Maezawa, who had also partially funded the BFR's development, heralded the sponsored dearMoon project. The mission consisted of the BFR crew with six to eight artists flying around the Moon while creating works of art in the duration. After the presentation, Musk then revealed the new addition of flaps to BFR: two at the top and three at the bottom, working as altitude controls during BFR's descent, with the bottom flaps doubling as landing legs. Starship was christened publicly in

November 2018, when the rocket booster was also designated as "Super Heavy". Starship has been designed to operate as a long-duration space vehicle on some missions or as the second stage of the combined Super Heavy/Starship two-stage rocket.

Structure and Technical Details

Structure

Starship is mainly constructed out of stainless steel, giving the rocket its signature silver gleam. The material was found to be more reliable. Rolls of stainless steel are delivered to SpaceX where they are then unrolled, cut, and welded along the cut edge to produce a nine-meter (30 ft) diameter cylinder, two meters (7 ft) in height, and weighing around 1,600 kg (4,000 lb). The welding is an innovative process called "friction stir welding", wherein a high-speed, spinning head of an automated, two-story welding machine practically punches along the two metal joins, fusing the crystalline structures of the metal pieces together. It makes the structure stronger without having to resort to using rivets or fasteners, thus saving mass, which is crucial to a spacecraft.

Starship's body is then made from 17 stacked, steel cylinders with a nose cone added and welded along their edges. Starship has four body flaps to control its orientation and falling velocity. Internally, domes separate the liquid methane and oxygen tanks at high pressure within Starship's body. The domes are manufactured by robots that weld the seams at a rate of 10 minutes per seam or four hours per dome. The full installation is then inspected with an X-ray machine.

And while Starship can gleam silver, on top of Super Heavy it also becomes the black tip of a space spear against the blue sky once its heat shield is applied. Installed are thousands of black, hexagonal tiles spaced out to counteract expansion at temperatures of 1,400 °C (2,600 °F), protecting Starship's skin from hot plasma damage. The heat shield is deemed to be reusable, theoretically requiring no maintenance between flights. However, SpaceX hopes the tiles' hexagonal shape will make it easier to be mass-produced.

Starship Spec

- Height: 50 m (164 ft) tall
- Diameter: 9 m / 30 ft
- Dry mass: Less than 100 tons (220,000 lb)
- Total propellant capacity: 1,200 tons (2,600,000 lb)
- Thrust: 1500 tf / 3.2 Mlbf
- Payload volume: 1,000 m3 (35,000 cu ft)
- Payload capacity: 100-150 tons (mission dependent)

The fuel tanks are divided into main and header tanks, the latter possessing more insulation and primarily used to flip and land the Starship following reentry.

As of August 2021, Starship's nose cone is created by two rows of stretch-formed steel.

Starship Variants

- **Crew**: One hundred people can be flown to the Moon, Mars, and other destinations, with Starship adapted to have private cabins, communal and storage areas, solar storm shelters, and a viewing gallery. Starship's life-support system is described as 'regenerative' by constantly recycling its resources.
- **Crew Lunar Lander**: Starship HLS is an adaptation for NASA's Artemis program. The lander may be installed with windows, air locks, an elevator, and landing thrusters. On an Artemis mission, Starship HLS may launch several months ahead of the crew and with a large amount of payload for setting up the lunar base. Starship tankers will refuel the HLS along the route to the Moon.

- **Commercial Lunar Payload Services**: HLS may also be used to send scientific, exploration, and commercial payloads to the Moon ahead of human missions.
- **Generic Cargo**: Spacecraft with a large door instead of normal payload fairings. Payloads, like satellites, can be mounted on the inside of the payload bay's sidewalls and released during space missions.
- **Tanker**: One Starship will refuel another Starship in orbit. In order to demonstrate this technology, NASA awarded SpaceX $53.2 million in October 2020 to perform a large-scale flight demonstration of transferring 10 tons (22,000 lb) of propellant between the two Starship tanks.

Super Heavy Booster Spec

- Height: 69 m / 230 ft
- Diameter: 9 m / 30 ft

- Dry mass: Between 160 tons (350,000 lb) and 200 tons (440,000 lb)
- Propellant: subcooled, liquid methane and liquid oxygen (CH4/LOX)
- Propellant capacity: 3400 tons / 6.8 Mlb or 3,600 tons (7.9 mlb) consisting of 2,800 tons (6.2 Mlb) LOX and 800 tons (1.8 Mlb) of CH4
- Fuel tanks weight: 80 tons (180,000 lb)
- Engine: Thirty-three sea level-optimized Raptors—more than twice as powerful as the Saturn V
- Thrust: 7590 tf / 17 Mlbf
- Gross liftoff mass: 3 million kg
- Super Heavy/Starship interstage weight: 20 tons (44,000 lb)
- Engines and mounts weight: 2 tons (4,400 lb)
- Grid fin weight: 3 tons (6,600 lb) per fin

Super Heavy Operations

Super Heavy has four grid fins (looking like large waffle-iron flaps) powered by electricity to control the booster's orientation and descent. Super Heavy controls many of its functions by firing cold gas thrusters using evaporated tank propellant. In order to separate from Starship in space, Super Heavy will shift its firing engines and release the latches. The grid fins circle the booster unevenly to allow it to attain more pitch control and to rotate in the roll axis. The booster will return to land at the launch site on its six legs, with the grid fins acting as touchdown-mounting points into the launch tower's recapture system.

Payload (Crew and Cargo) Spec

Starship's payload possesses the largest usable payload volume of any current spacecraft and is larger than the ISS pressurized volume. This payload space can be utilized for both crew and cargo.

- Payload volume height: 18 m / 59 ft
- Payload fairing diameter: 9 m / 30 ft
- Payload volume: 1,100 m3 / 38,800 ft3
- Useful carrying mass: 100+ tons / 220+ klb

Raptor Engines

Starship's Raptor engine is a reusable, methalox, staged-combustion engine designed and manufactured by SpaceX. Starship has six Raptor engines, three for lower atmosphere operations and the other three Raptor Vacuum engines designed for the vacuum of space. SpaceX's Raptor engine family is the only existing full-flow, staged-combustion cycle engine, a feat even the United States and Soviet Union failed to accomplish for a rocket.

A full-flow, staged-combustion cycle engine generally has two preburners attached to their corresponding turbopumps. While one of the preburners is supplied with an oxidizer-rich mixture, the other has a propellant-rich mixture. The preburners ignite their mixtures marginally to drive their corresponding turbines, then the cycle feeds all of the oxidizer-rich and propellant-rich mixture into the main combustion chamber, increasing the engine's chamber pressure to about 300 bar (4,400 psi), the highest of all current engines, without wasting any propellant, unlike other engine processes. The overall efficiency is increased, creating more thrust.

SpaceX began flight testing their Raptor engines in July 2019, achieving the first ever flight test of a full-flow, staged-combustion rocket engine. SpaceX hopes in the future to streamline their Raptor manufacturing so that the engine may be mass-produced and bring down costs to about $230,000 per engine and $100 per kilonewton.

Raptor Engine Spec

- Material: New alloy (dubbed 'SX500' by the SpaceX metallurgy team)

- Diameter: 1.3 m / 4 ft
- Height: 3.1 m / 10.2 ft
- Propellant: Subcooled liquid methane and liquid oxygen (CH4/LOX)
- Thrust: 230 tf / 500 klbf—each launch enough to send payloads of at least 100 tons / 220,000 lb into low-Earth orbit.

SpaceX manufactures variants of Raptor for varying purposes.

- Raptor Vacuum will be the space variant, increasing the specific impulse or fuel efficiency to 380 seconds.
- Raptor 2 is the next generation in the family, producing more thrust and has its specific impulse reduced by three seconds.
- Sea level-optimized Raptor engine with gimbaled thrust.
- Sea level-optimized Raptor engine without gimbaled thrust.
- Vacuum-optimized Raptor engine without gimbaled thrust.

Fuel

Methane was selected for the Raptor engines for its cheaper cost, noncoking attributes, and most importantly, for Mars missions. Methane can be produced on Mars through the Sabatier reaction. In 2019, SpaceX published their Raptor exhaust plume calculations. Carbon dioxide, water, trace amounts of carbon monoxide, and nitric oxide (more on that later in Chapter 9) were expected to be present.

The versatility of Starship and Super Heavy is clearly demonstrated through Starship's constant iterative advancements, which creates a flexible yet deliberate, inventive, and technological process. The path toward Starship has been a 20-year journey of pushing boundaries traditional space organizations never knew existed, let alone thought to cross. And though Starship has gone through several iterations and configurations, it now sits on a majestic pedestal waiting to forge a new future for humanity.

Chapter 3:

Starship Testing

"My vision is for a fully reusable rocket transport system between Earth and Mars that is able to re-fuel on Mars - this is very important - so you don't have to carry the return fuel when you go there."

-Elon Musk

If you think back to the Saturn V Apollo launches, they consisted of a two-stage rocket with a capsule on top. Starship is a two-stage rocket, but SpaceX has turned the second stage into its own spacecraft and a reusable one once it separates from Super Heavy. This is another technological innovation that has garnered SpaceX plaudits in the space community. This allows Starship to transport more people than ever before into space. But first, the systems have to be tested.

Between July 2019 and May 2021, eight prototypes of the Starship upper stage, each with different vehicle iterations, have been flown nine times: Starhopper, SN5, SN6, SN8, SN9, SN10, SN11, and SN15. All the development test flights are launched from the Boca Chica Starbase. SpaceX tests rocket prototypes, ranging from proof pressure tests and static fires to full-flight tests with attempted recovery. Due to SpaceX's transparency with space news outlets, the tests are covered extensively.

While the public can see the tests of each Starship iteration live or through pre-recorded videos, SpaceX rocket tests are proprietary in that no "detailed set of test objectives" are released. A date is set and publicized, the test undertaken, and though some of the rockets have not survived the testing, the actual objective may have been something more mundane in line with the iterative policy of SpaceX. Each test result just has to exceed the last one, qualifying the data for future tests.

The first testing of the Raptor liquid methane engine components, still intended for the pre-Starship Interplanetary Transport System (ITS), was undertaken at NASA's Stennis Space Center. To aid in this collaboration, SpaceX installed more equipment to the existing infrastructure. From 2014 to 2016, the Raptor engine was based at the Stennis E-2 test complex with hopes of generating more than 2,900 kN (661,000 lbf) vacuum thrust. During this time, multiple main injector testing and full-power tests of the oxygen preburner component were also completed.

Early 2016 saw SpaceX completing construction of a new Raptor engine test stand at their McGregor test site in Central Texas. By August that same year, it had undertaken delivery of the first integrated Raptor rocket engine for testing. The engine had been manufactured at SpaceX's Hawthorne facility and was the first ever full-flow, staged-combustion, methalox engine to reach a test stand. On September 26, 2016, SpaceX performed an initial, nine-second firing test. The engine had one MN (220,000 lbf) of thrust, which was less than half the thrust of the eventual full-scale Raptor engine in Stennis. However, the test engine was only one-third the size of succeeding engine designs to follow.

By September 2017, the Raptor engine testing regime had claimed a cumulative 1,200 seconds of test-stand fire testing in the course of 42 main engine tests. The longest test lasted 100 seconds, but was limited by the capacity of the ground-test propellant tanks. To enhance the performance, SpaceX engineers created a new alloy they called 'SX500' to protect the Raptor. The alloy will assist its oxygen-rich turbopump in resisting oxidation, as the injected hot oxygen gas in the engine would be in excess of 12,000 psi.

In the Starship era, six Raptor engines will power Starship (three sea-level Raptor variants and three vacuum variants). The cumulative thrust of these engines will be about 14 MN (1,400 tf; 3,100,000 lbf). Super Heavy will require 33 sea-level, variant Raptor engines. At the very bottom of Super Heavy is the engine housing, known as the 'skirt'. Within the skirt are also helium gas pressure vessels which spin up the Raptor turbopumps. Above the skirt are the liquid oxygen and liquid methane propellant tanks, with a methane header tank separating them. This methane tank contains Starship's propellant for landing.

In July 2019, Starhopper became the first prototype to fly using a single Raptor engine. The stainless-steel test platform had three non retractable, extended fins for legs and measured about 20 meters tall and nine meters across. All the launch test flights were at Boca Chica, Texas. The first three tests were tethered, but the last one was untethered, and Starhopper also traversed laterally before touching down on a landing pad nearby.

Test Date	Test Height	Test Duration
3 April 2019	<0.3 m (1 ft)	~3 seconds
5 April 2019	1 m (3.3 ft)	~5 seconds
25 July 2019	25 m (80 ft)	~22 seconds
27 August 2019	150 m (492 ft)	~1 minute

Starhopper has since been retired and repurposed as a water tank with supporting mounts for communication, weather monitoring, and tracking equipment. It sits close to the test launch site.

After Starhopper's successful tests, Starship upper-stage prototypes Mk1 and Mk2 were constructed. Mk1 was located in Boca Chica, Texas, while Mk2 was held at SpaceX's facility in Cocoa, Florida. During this development, in late September 2019, Musk updated details on Starship's lower-stage booster (Super Heavy), the upper stage's descent control system, its heat shield, orbital refueling capacity, and potential destinations outside Mars. It was at this time Musk announced Starship's material had been changed to stainless steel from carbon composites. Stainless steel provided lower costs, a higher melting point, greater strength at cryogenic temperatures, and better manufacturing prospects.

Between SN4 and SN5, during June 2020, SpaceX accelerated the construction of infrastructure at the spaceport, including the launch pad for orbit-capable Starship rockets. Large tents, station facilities, and repurposed freight containers became the backbone of their production line, enhancing rocket construction. All subsequent flight tests were at Boca Chica.

- November 20, 2019, Mk1 was destroyed during a cryogenic pressure stress test
- Mk2 was not flight tested, as the Florida facility was being deconstructed throughout 2020
- After Mk1 and Mk2, SpaceX began designating Starship's upper-stage prototypes with the prefix 'SN'

Test Flight No.	Test Date	Test Flight Height	Flight Test Duration
SN1	28 February 2020	N/A—collapsed during pressure stress tests.	N/A
SN2	8 March 2020	N/A—passed the pressure test.	N/A
SN3	3 April 2020	N/A—collapsed during firing tests.	N/A
SN4	29 May 2020	N/A—exploded after its fifth engine firing.	N/A

SN5	4 August 2020	150 m (492 ft)	~45 seconds
Prototype SN5 lacked flaps and a nose cone, so it had a cylindrical shape. It possessed one Raptor engine, propellant tanks, and a mass on top.			
SN6	3 September 2020	150 m (492 ft)	~45 seconds

SN7, SN7.1, and SN7.2 were practice vehicles testing different grades of the stainless-steel thickness in construction. They underwent cryogenic and pressure testing to eventual destruction, with SN7.2 being repaired and retired.

High-Altitude Flight Tests (Starship SN8 to SN15)

Test Flight No.	Test Date	Test Flight Height	Flight Test Duration
SN8	9 December 2020	12.5 km (41,000 ft)	6 minutes, 42 seconds
SN9	2 February 2021	10 km (32,800 ft)	6 minutes, 26 seconds
SN10	3 March 2021	10 km (32,800 ft)	6 minutes, 24 seconds

SN11	30 March 2021	10 km (32,800 ft)	~6 minutes
Starship prototypes SN12, SN13, and SN14 were scrapped before completion.			
SN15	5 May 2021	10 km (32,800 ft)	5 minutes, 59 seconds
SN16-SN19	N/A - scrapped	N/A	0.00

SN8 was the first complete Starship upper-stage prototype. It performed four static fire tests between October and November 2020. On December 9, 2020, SN8 ascended to an altitude of 12.5 km (7.8 mi) using three Raptor engines and performed its landing flip ("belly flop") maneuver using its flap controls, before reigniting the engine for a vertical descent. However, on its landing approach, a few seconds before touchdown, a pressure fault in the methane fuel tank caused the craft to lose thrust and impact the pad, destroying SN8.

Starship SN9

On February 2, 2021, SN9 launched to 10 km (6.2 mi). SN9 also had three Raptor engines allowing it to successfully execute its propellant transition to the internal header tanks holding its landing propellant. SN9 "belly flopped" as the engines turned off, before their planned re-ignition and reorientation to vertical with use of all four flaps. However, during the landing flip maneuver, one of the Raptor engines did not relight, causing SN9 to over rotate and crash into the landing pad, or as SpaceX stated, SN9 experienced a 'RUD' (rapid unscheduled disassembly).

Starship SN10

On March 3, 2021, SN10 launched on the same flight path as its two predecessors, hitting 10 km in altitude. Propellant transition was accomplished, engines shut down and reignited, and aerodynamic descent started. The landing was successful, but very hard, crushing the landing legs and causing a lean to one side and fire. Eight minutes after landing, most likely due to a ruptured propellant tank, SN10 exploded.

Starship SN11

Due to thick-fog conditions, SpaceX gained approval from the FAA to launch SN11 on March 30, 2021. It flew along the same flight path as its forerunners. However, upon engine reignition, one of the Raptors exploded at an altitude of around 600 meters. Debris was scattered up to 8 km (5 mi) away.

Starship SN15

On May 5, 2021, the upgraded SN15 launched in overcast weather. As per its predecessors, it accomplished the same engine and flap control maneuvers, flipped over, descended, and touched down softly on the landing pad. It was the first successful Starship prototype high-altitude test. A small fire did start near the base shortly after landing, but it was out within 20 minutes.

As of July 2021, the planned tests and launches of SN16 to SN19 have been scrapped and canceled. SN20 was the next Starship prototype awaiting a suborbital flight with a Super Heavy booster. By August 2021, the iterative testing process was now ready for the first orbital test flight of the two-stage Starship system.

Super Heavy Testing

Super Heavy has undergone static tests under the designation of 'BN', akin to Starship's SN. Below is a summary of construction and test results.

Super Heavy Ref No.	Static Test Date	Status	Flights
BN1	N/A	Scrapped	0
BN2.1	8 June/17 June 2021	Cryogenic tests	0
B2.1	1-3 December 2021	3x Cryogenic tests	0
BN3	19 July 2021	Scrapped	0
B4	N/A	Retired	0
B5	N/A	Retired	0

B6	Not Yet	Converted into test tank	0
B7	Not yet	Built	0
B8	Not Yet	Under construction	0

Super Heavy BN1

The first Super Heavy Booster prototype was designed as the guide reference for the following builds and not for flight testing. It was 70 m (230 ft) tall, and between December 2020 and March 18, 2021, it had been constructed and stacked inside Boca Chica's high bay building, before being scrapped in April 2021.

Super Heavy BN2.1

BN2.1 was rolled out to the launch pad on June 3, 2021, for successful cryogenic tests on June 8 and June 17, 2021, then subsequently retired.

Super Heavy B2.1

B2.1 was another booster test tank making three cryogenic tests on the 1st, 2nd, and 3rd of December 2021. On December 6, 2021, it was rolled back to the production building and retired.

Super Heavy BN3/B3

BN3 or "Booster 3" was anticipated to be the first booster to make an orbital flight, but was only used for ground tests. A cryogenic proof test was performed on July 13, 2021. It was rolled out to the test pad July 1, 2021, for static engine testing. Three Raptor engines were installed, and a live fire test was conducted on July 19, 2021. A further nine Raptors were due to be fitted; however, BN3/Booster 3 was partially scrapped on August 15, 2021, with a full removal and scrap on January 9, 2022.

Super Heavy B4

Elon Musk had several hundred SpaceX's employees transferred from Hawthorne to Boca Chica to accelerate the development of SN20 and BN4 for August 2021's rollout. It appeared BN4 and SN20 would be the first fully-stacked Starship on the pad ready for the first orbital flight test. The plan was to have Starship launch for a suborbital flight with BN4 performing a soft water landing in the Gulf of Mexico. BN4 had 29 Raptor engines installed—four less than the original design for 33 engines. Grid fins were fitted for atmospheric reentry testing.

SN20 was initially stacked on BN4 on August 6, 2021 for a fitting test. Then, Starship was disassembled for BN4 to be rolled back to the high bay for secondary wiring. On September 9, 2021, BN4 was rolled out to the launch site and mounted on the newly-modified Orbital Launch platform.

There, it completed cryogenic and pneumatic proof tests between December 17 and 22, 2021, before being retired on August 14, 2022.

Super Heavy BN5

Construction of B5 was seen in July 2021. It was finally rolled out on December 8, 2021 and placed on a display stand in Boca Chica's "Rocket Display Garden" alongside SN15 and SN16 for display or possible later scrapping.

Super Heavy B6

Construction of B6 was seen in August 2021, but by December 8, 2021, it was converted into a test tank.

Super Heavy B7

B7 was under construction as of September 2021. It was rolled out to the launch site on March 31, 2022 and first placed on the orbital launch mount, where it completed a cryogenic proof test on April 4, 2022. It was then moved to the new booster test stand on April 8, 2022, where it completed another cryogenic test on April 14, 2022. However, the booster suffered a failure and ruptured, so repairs had to be made.

On May 5, 2022, B7 was rolled back to the launch site and completed two cryogenic tests on May 9 and 11, 2022. More repairs and upgrades were completed in May and June 2022 in Boca Chica's new "Mega Bay" (aka "Wide Bay" or "Highbay 2"). As of June 18, 2022, B7 has been returned to the launch site to undergo its static testing operations.

Super Heavy B8

Construction for B8 was seen in January 2022 and is still under wraps.

Planned Orbital Flights

The combined Starship booster and upper stage are yet to fly, but SpaceX has announced several updates on developments and their plans for Starbase and future flights. A new Starship was unveiled with the designation of 'SN20'. The complete spacecraft has not flown yet.

Starship + Super Heavy

Date and time	Flight apogee height	Duration
(NET) "no earlier than" June 2022	(TBD) "to be determined"	~90 minutes (planned)

In March 2022, Elon Musk announced that "new prototypes will conduct the orbital flight", not SN20 and BN4. Whenever the test proceeds, the mission plan is for Super Heavy to separate from Starship around three minutes after launch and splash down about 30 km (19 mi) offshore in the Gulf of Mexico. Starship will then continue its flight over the Florida Strait, avoiding populated areas while accelerating to orbital velocity. It will then reenter the atmosphere over the Pacific and perform a soft splashdown about 100 km (62 mi) northwest of Kauai, Hawaii. It is not known when Starship will take its first suborbital steps.

Further developments have also been observed:

- SN21 has been scrapped.
- SN22 retired and moved to the Rocket Display Garden in February 2022.
- SN23 has been scrapped, with its parts being used in SN24.
- SN24 has continued to be stacked.

- SN25 is under construction as of February 2022.
- SN26 may also be under construction as of February 2022.

A Path Towards Mars

A source of delays to Starship's orbital testing was the long-awaited Federal Aviation Administration's (FAA) release of its final Programmatic Environmental Assessment on June 13, 2022. This outlined SpaceX's Starship flight parameters, which had been sent to the Federal Communications Commission (FCC), and also assessed the environmental impact of the subsequent testing and flights.

When the report was released after a year and a half of determinations, the FAA approved SpaceX's Starship launch plans from Boca Chica. The FAA's environmental review had concluded

SpaceX's plans for orbital launches posed "no significant impact" throughout the Gulf Coast region.

Though there are 75 items of action SpaceX must complete to lessen the ecological impact to the surrounding environment by Starship launches, in effect, it is a green light for Starship to launch sooner rather than later. Actions SpaceX must take include giving earlier notice of its launches, coordinating with biologists and state and federal authorities to monitor the effects of launches on wildlife and vegetation, remove all launch debris from sensitive habitats, and reduce their lighting in the region to minimize its impact on wildlife. SpaceX must also limit its closures of a nearby highway so members of the public can still visit the nearby beach, specifically on 18-agreed holidays, and the highway/beach can be closed no more than five weekends annually.

Even though SpaceX has approval for launch, it still has to acquire a license from the FAA to *actually* launch, which could see more regulatory checks, legal obstacles, and delays. Other delays could also accrue from legal disputes with local people, other US agencies, and other aerospace companies (such as Blue Origin) over NASA contracts. However, despite this, Elon Musk, who is well-known for his overly-ambitious schedules, has stated Starship will be "ready to fly" into Earth's orbit next month.

SpaceX has been ready for this through its iterative testing programs—building, testing, crashing, and recovering rockets. Unlike traditional space agencies, SpaceX has several prototypes under construction at the same time to be ready to go, whether the previous test was a success or not. It's all part of their plan. In fact, Musk, again, has stated they "have a second Starship stack ready to fly in August and then monthly thereafter." Currently, SpaceX has a full-size, Super Heavy booster awaiting testing and currently has five spacecraft in production. It's a master plan that will surely get humanity to Mars.

Chapter 4:

Diversion to the Moon

Before Musk goes to Mars, the Moon will be SpaceX's first destination on its multiplanetary mission. Starship's systems, engines, tankers, crews, and mission process would have to be tested much closer to Earth, but farther out than the ISS.

Setting up colonies on the Moon would be a priority for both NASA and SpaceX to validate Starship, to fulfill NASA's promise of returning to the Moon, and to galvanize the world for a Mars mission. Though there are disagreements within the space industry that Moon missions are a prerequisite for a Mars mission, the fact is SpaceX *has* NASA contracts to fly Starship to the Moon.

First to the Moon

"We choose to go to the Moon in this decade and do the other things, not because they are easy, but because they are hard…" So spoke President John F. Kennedy in 1962, outlining the ambitious goal of the US space program.

But by 1972, the Moon race was over and only 12 men have ever set foot on the Moon. Since then, there have been ample expensive Moon programs lofted forward and empty words said and promised. NASA had no concrete plan or infrastructure to return humans to the Moon.

Then, in 2017, NASA's Artemis program to launch astronauts to the Moon was established, with the Space Launch System (SLS) carrying the crewed Orion vehicle to the Moon. It seemed a promising and viable plan.

However, with heavily-delayed development plans, costs spiraling, and with SpaceX's Falcon and Dragon successes with NASA on supply and crew missions to the ISS, the space agency selected SpaceX's variant Starship HLS as the crewed lunar lander on April 16, 2021. There was a legal dispute with fellow contract tenderers, Dynetics and Blue Origin, over the decision, but the last legal complaints were dismissed by the Court of Federal Claims three months later. SpaceX is now free to conduct work for NASA and to take on other commercial ventures, as summarized below.

Starship and NASA to the Moon

Artemis

One of Starship's first tests to make humanity a multiplanetary civilization begins with the Moon. In the proposed scenario, under NASA's contract, SpaceX will develop, construct, and fly two lunar landing flights.

For the Moon mission, there are a lot of moving pieces involving NASA's SLS and Orion spacecraft, Gateway (a new, Moon-orbiting space station), and Starship HLS and tankers. Once launched, Starship HLS will refuel in space for its voyage to the Moon. Once there, HLS will then dock either with Gateway or directly with a crewed Orion spacecraft, which has launched from Earth via the SLS. HLS will then land that crew with supplies, equipment, and science payloads onto the Moon for days of lunar operations. When the surface objectives are complete, HLS will launch and dock with Orion transferring the crew back onto itself. Orion will then return to Earth.

Starship HLS will then stay on station, docked at Gateway. Here, HLS will be refueled by a Starship tanker and await the next Orion mission for the Moon.

DearMoon Mission

As previously mentioned in Chapter 2, in September 2018, Japanese billionaire Yusaku Maezawa announced an extraordinary, combined touristic art project around the Moon. He, eight artists, and two crew members would become some of the first humans to travel aboard Starship, as it is launched on a trajectory around the Moon and back. The lunar journey is expected to take a week with a launch date provisionally set for 2023.

For the proposed journey, Starship and Super Heavy (acting as the booster) will lift off. After burning for around three minutes, Super Heavy will separate from Starship and return to land back on Earth while Starship continues into orbit. To accommodate the passengers, Starship will have its payload area pressurized to 1,000 m3 (35,000 cu ft), to provide safe space for large common areas, central storage, a galley, and a solar storm shelter.

Starship will then carry out a lunar-transfer burn using the three Raptor vacuum engines to get its trans-lunar injection lined up correctly for close approach and lunar flyby. Soon after, the Moon's surface will loom large in Starship's windows as it enters Perilune, the closest point to the Moon. And all along the way, the artists will be creating their masterpieces from their experience and vision of the Moon.

Once the flyby mission is over, the Starship will return to Earth, performing a final series of control burns to ensure a vertical alignment before landing on the touchdown pad. The works of art would be displayed around the world to promote peace. As of September 2022, Maezawa does not have a full roster of invited artists, yet.

Though Musk is more interested in Mars, going to the Moon demonstrates SpaceX's innovation and flexibility in its spacecraft and multiplanetary goals. If SpaceX can be successful in both the NASA and dearMoon missions, then its one small step to the Moon will all but ensure its giant leap to Mars. It is a big 'if', as Starship's development test dates keep slipping, and 2023 may be too soon to gamble on a Moon folly. Still, the race to a manned Mars mission is on, and it seems there is no one else in the running except SpaceX. In fact, you can almost hear Elon Musk state, "We choose to go to Mars, and do the other things, not because they are easy and not because they are hard, but because they are the right things to do, to make humanity a multiplanetary species."

Chapter 5:

Starship's Mission to Mars

"**M**ars within our lifetime." That has always been the mantra from space agencies and politicians, that we were only five, 10, 15 years away from going to Mars. But the will of politicians to spend billions getting to Mars during their scant few years in office, only to see their successor take the glory or for space funds to be perceived as a waste by the public when there are more pressing problems on Earth, would be too much for a politician to gamble their career over. And if one takes into consideration the budget required for space agencies like NASA or Roscosmos to get to Mars—hence the lower-cost robotic missions—then Mars is a dream too far out in our lifetime.

It was always suspected that Mars would be conquered by a private-public enterprise. The cost to get to Mars would be prohibitive for a single state to handle. So, the private organization, whether space-orientated or not, would sponsor a nation's stride to the Red Planet with commercial rights from mining or exploration going to the private sector and the public entity undertaking scientific endeavors for humanity. The other route to Mars was with a joint-nation mission for humanity with the US, Russia, China, ESA, and other national space agencies chipping in to fund, share space vehicles and crews, and travel to Mars for scientific cooperation and peace. But apart from the ISS and NASA/ESA missions, cooperative space operations seem to be more nationalistic than ever. Then, two things occurred to make a Mars mission more probable.

First, in 1998, a small group called the "Mars Society" started to advocate for a Mars mission. Its founder, Robert Zubrin, had the insightful idea to treat Mars like the early American frontier, when the first explorers lived off the land without having to carry all their resources from home with them. Mars' regolith (soil) and atmosphere was the raw material astronauts would need to make enough air, water, and fuel mixed with a small amount of human-brought materials.

The theory was that roughly every 26 months, when Mars is in conjunction with Earth (furthest away from Earth, but easier to travel to Mars), an unmanned rocket would be sent to the Red Planet. This rocket would have tons of hydrogen in storage, a chemical plant, and a small nuclear reactor on board. Once on Mars, after its six-month trip, the chemical plant would then mix the hydrogen with Mars' carbon dioxide atmosphere through the Sabatier reaction and electrolysis to transform the elements into tons of methane and oxygen. Over the course of around 10 months, less than 10 tons of hydrogen from Earth would create more than 100 tons of fuel and oxygen (with the hydrogen further mixed with the oxygen to produce water) on Mars. This rocket and chemical plant would then wait for the manned mission.

Then, if the fuel and oxygen production was a success with confirmation radioed to Earth, 26 months after the unmanned launch, a manned spacecraft with a "Habitat Unit" would be launched to Mars. Thus, within a four-year period, the astronauts have the means to survive and return home already on Mars. And every 26 months, the process would repeat, and small colonies could be built.

It was a far cheaper and more elegant plan than President George H. W. Bush's $500 billion (over 20 to 30 years) Space Exploration Initiative (SEI) to build a massive infrastructure of space stations, spacecraft, and colonies on the Moon and Mars. By contrast, the Mars Society's Mars Direct plan was estimated at $55 billion over 10 years. It was therefore not surprising when NASA approved a Mars Semi-Direct plan which replaced the SEI. But still, we are no closer to getting to Mars. Then came SpaceX.

MarsX

After the mockery and hostility SpaceX had first faced when starting out, especially with Elon Musk's future vision for Mars and humanity, most space industry insiders see SpaceX as the best chance of getting to Mars. Even with Musk's overly-optimistic timeline for a Mars mission, SpaceX has transcended all expectations of success. However, while Musk has a plan to send one million people to Mars by 2050, he and SpaceX have seemingly not addressed the hardest part of the mission. We know how Starship will launch, and we know the destination, but the middle bit—the *actual journey* to Mars—is almost a complete blank.

Unlike Zubrin's Mars Direct plan, SpaceX has not outlined details for life-support systems, radiation protection, and in-situ resource management while on the Red Planet, essential for a successful Mars mission. How will SpaceX deal with months of weightlessness, physical and mental wellbeing, boredom, loneliness, and also the wealth and brain drain as the best and brightest leave Earth for Mars?

Mars Direct and SpaceX

Having interacted before in the Mars Society, Robert Zubrin believes SpaceX's heavy lift capability and low-cost launch options will herald lower-cost missions to Mars. A Mars Direct mission would be attainable by force of SpaceX's rocket power and possessing more living space for the astronauts on the journey. Of course, Zubrin, in referring to Falcon Heavy and Dragon capsules, had not envisioned Starship at the time and plans for 100 people to travel to Mars at a time, but crews of three to six, at most. Musk still optimistically predicts SpaceX will land humans on Mars before 2029 (coincidentally 60 years after Apollo 11 landed on the Moon). In order for this to happen and to mitigate less-than-optimal outcomes, it may be advantageous for

SpaceX and the Mars Society to collaborate on the human safety aspects of the mission.

Funding the Mars Mission

Starship's mission to Mars is financially supported by only SpaceX and Elon Musk's personal capital. Much of what SpaceX is doing in progressive development and manufacture of space technology is so innovative and 'multifaceted' it has reduced launch costs so that a Mars mission is feasible within a decade.

But as the Washington Post had reported, "The [US] government doesn't have the budget for Mars colonization. Thus, the private sector would have to see Mars as an attractive business environment. Musk is willing to pour his wealth into the project." But it is argued if this will be enough to create the colony he envisions.

According to sources from October 2016, SpaceX was expending "a few tens of millions of dollars annually on development of the Mars transport concept, which amounts to well under five percent of the company's total expenses." By 2018, expected costs are projected to rise to around $300 million USD per year. It is estimated that overall costs to create the infrastructure around Mars launches will be around $10 billion USD, with SpaceX looking to even more before it generates any revenue from the Mars missions.

However, contrary to some industry reports, Musk stated in September 2016, and repeated in October 2017, that he foresees Mars colonization plans being financed by both private and public funds. Essentially, SpaceX would provide the commercially-available Mars transport according to demand, the propellant production infrastructure, and development funding. The establishment of a base would be managed by other third-party companies and organizations.

For people traveling to Mars, Musk suggested there could be means to take out a loan to pay for their trip. In order to pay off the debt, people might work on Mars to reduce the anticipated labor shortages on the Red Planet. Though no current infrastructure seems to be set up for this eventuality, one such mutually beneficial plan could be for mining companies to sponsor miners on Mars and pay off their debt by discovering and claiming mineral-rich lands for them. Whatever the plan, those involved will have to ensure funding is in place for travel and that Mars is equipped for their survival.

Criteria for Going to Mars

Both the Mars Society and Elon Musk have echoed the famed words attributed to Antarctic explorer Ernest Shackleton, who was recruiting men for one of his expeditions: "Men wanted for hazardous journey. Small wages, bitter cold, long months of complete darkness, constant danger, safe return doubtful. Honor and recognition in case of success."

The Mars Society has used a similar advertisement to enlist volunteers for their Mars Desert Research Station (MDRS) in Utah and the Flashline Mars Arctic Research Station (FMARS) in Devon Island, Canada. Elon Musk has likened Starship's Mars mission in the same adventurous vein, but also sees far more opportunity. He does not see Mars as an "escape hatch for rich people", but he can see that many of the one million people who go to Mars will have "an abundance of entrepreneurial activity, because Mars will need everything from iron foundries to pizza joints."

For himself, however, Musk acknowledged, in 2013, he would only personally travel to Mars if he knew SpaceX could continue running without him. He added, "I've said I want to die on Mars, just not on impact." Perhaps, once one million people have made it to Mars, Musk will feel more confident in SpaceX's survival and maybe even feel obligated to go to Mars. Certainly on Mars, Musk's entrepreneurial spirit could push even more innovative ideas and visions into play.

Training for Mars

As it is, SpaceX has started its own astronaut training program seeking to hire Space Operations Training Engineers. NASA's process of training astronauts takes around two years. There is expected to be cross-training between SpaceX and NASA, with the established physical and psychological tests to ensure the candidate spacefarers have "The Right Stuff". But how will civilians be trained? How do you thoroughly vet 100 passengers from across the globe per 1,000 flights? Not all the volunteers for Mars will be engineers or technologically savvy. Not all will have physical training levels of elite astronauts, being all different ages with varying life experiences.

In fact, how would you continuously train a million people to go to Mars? The mission will be hard-going, dangerous, breath-taking, and life-affirming. But the facility to train and evaluate potential Mars travelers seems to be absent. Initial online questionnaires may get you in the door, but the expected thousands who would apply for the first few flights would surely overwhelm the system. Deciding on final candidates and training them could take longer than an open-Mars travel window, delaying the mission by years. But so far, no plan has been announced.

NASA would most likely train the professional crews of pilots, engineers, scientists, and mission specialists, but what about those with little-to-no technical skills? The Mars Society has been running their Mars analogue research stations for years in Utah and Devon Island, Canada. There had been plans for a Euro-Mars hub in Iceland and Mars-Oz in Australia, but these dormant plans require funding. However, there may be a promising start to a new Mars research station in Mongolia's Gobi Desert, started by the Mongolian Mars Society. What if SpaceX were to collaborate with the Mars Society and other national space agencies to fund these and other start-up research habs around the world? That would create a global training program run by professional organizations, and for every Starship mission within a short span of time, SpaceX would have well-trained people ready for Mars.

Why Go to Mars?

It's the million-dollar question that will be asked of the one million people who may go to Mars: "Why do you want to go to Mars?" There's no fresh air, no greenery, cramped spaces, dangerous terrain, and the chance you might not return (if that was your plan). Just going to Mars to get away from Earth may not cut it as a valid reason to go. You would need some skills or a means to contribute to the colony until it was self-sufficient enough to accept tourists, non skilled colonists, and other people.

But there are reasons people would go to Mars. Elon Musk has already let it be known that Mars would be the start of humanity's multiplanetary journey, assuring that a vestige of the human race would survive "out there" should something happen to Earth. Humanity needs the challenge to survive, or we will grow corrupted and/or complacent, dying off without accomplishing anything in the cosmos. One day, Earth could become the bottleneck causing humanity's demise. It would be prudent to have the safety valve of another world to carry on humanity's story.

Discoveries would be another incentive for people to go to Mars, whether the search for life, new Martian water supplies, exploration of lava tubes and other unexplored features, or expeditions up Olympus Mons. But the search for life would be uppermost learning if an independent ecosystem lived or still lives hidden on Mars or if Martian life kickstarted life on Earth through Martian meteorites landing on the planet with microorganisms within them. The question "Are we alone in the universe?" could be potentially answered on the Red Planet.

Such discoveries on Mars could lead to spin-off technologies which could be useful to Earth. Even venturing out to the asteroids to mine them could net a fortune for a private company, but provide vital resources back on Earth. Less mining on Earth could start a new environmental impetus to combat climate change.

Space will be a valid employment industry with jobs on Earth, in low-Earth orbit, the Moon, space stations, and on Mars. It would create a new generation of spacefarers with high STEM skills ready for the next century of human space exploration. Space should also promote peace and cooperation between nations. It is vast with room enough for all, even Mars, which has the same surface area as Earth's continents. Mars wouldn't become as overcrowded as Earth, at least not in the near future.

There would be scope for space tourists to go to Mars. They've made a lot of money, traveled the Earth, and sought a new adventure in space. Perhaps they'll settle on Mars for the thrill of it, start a new business there, or keep on trucking across the Martian plains.

So, you have cited your reasons for going to Mars, made it through the vetting process, and passed the physical and psychological tests. There may still be further tests to assess your reactions to boredom, loneliness, and homesickness, even cooped up with 100 other people on a confined spaceship for six to eight months. There are bound to be personality clashes amongst strangers, cultural differences, and conflicting philosophies on religion, politics, and how to live on Mars. It should be assumed SpaceX will choose an international manifest for Starship so that the first crews on Mars won't be all American citizens. Would Russians or Chinese be excluded from participating? Elon Musk's multiplanetary vision would have to include as many ethnicities as possible, Starship being the Noah's Ark of space. But the passengers will be human, after all, and unknowingly take their biases with them. This is where some form of health and security management would have to be implemented to monitor the passengers and crew for any dissenters, sending updates back to Earth for evaluation.

As well as a medical bay, it would do well for Starship to also have a brig.

The Wealth and Brain Drain

In 2012, Musk stated that the price guide for a Mars voyager would be around $500,000 USD per person. In 2016, he rounded this down to as low as $200,000 USD and has further implied a ticket price for about the median price of a home in the United States. So, relocation to Mars could involve selling one's home rather than getting a loan. These costs are a lot less than the tens of millions paid by "space tourist" billionaires to the ISS. But as there will be a 100 people onboard Starship, this would still generate around the same amount of fees paid by the billionaires per ISS flight. As a commercial company, with a potential monopoly of manned Mars missions for years to come, SpaceX could charge what they wanted, but not enough to deter the million people they need for Mars.

There will be arguments that such costs are exorbitant for the ordinary person. It may be an option to take out a loan for a Mars mission (though such loans may be easier with the establishment of the Bank of Mars). So, of course, it will be the wealthy and talented who would leave Earth, creating a potential wealth gap and brain-drain scenario. They would be people who got rich through innovations and those who built the world, the engineers and scientists seeking fame and fortune in their respective fields. Perhaps their discoveries and inventions on Mars would be applicable on Earth, or perhaps the wealth and brain-drain gaps would be filled by others on Earth. As the saying goes, "Nature abhors a vacuum," so if the best and brightest left Earth, then like natural selection, those niches would be filled in compensation. Humanity would survive.

Mars could be an insurance policy for humanity on Earth, but insurance for Mars trips would also soar. The space industry would provide the insurance trade with new revenue. New financial institutions just dealing with Mars would pop up, whether as brick-and-mortar, online, or crypto businesses. And it would not just be those industries to benefit. Communications, medicine, transport, manufacturing, etc., could be transformed. One million people 'missing' from the world would be a drop in the population ocean, but Starship will not just be transporting people to Mars for a new life. When those who are left behind look up to the stars, they will know a small part of humankind is out there, inspiring others to lead and to dream, opening up new opportunities, and daring them to follow.

Starship's Journey to Mars

In October 2020, Elon Musk stated an uncrewed mission would head for Mars in 2024 (since revised to 2029). Musk's plan is to have 1,000 Starships launched to Mars every 26 months, leading to the development of a sustainable Martian city of one million people within 50 to 100 years. The journey to Mars will take around 80 to 150 days. The first of the Starship missions would have smaller crew sizes, 12 or so, and that way, more of the pressurized payload space could be used for cargo. The crew's goal would be to construct and troubleshoot issues with the base and the propellant plant power system. In the event of an emergency, Starship would be able to return to Earth using the produced propellant. But there would be a couple of problems to overcome before Starship was to launch to Mars: radiation and weightlessness.

Radiation Shielding

Once on Mars, its thin atmosphere and built-in or natural shelters would shield humans from harmful radiation. However, there has always been cause for concern about radiation levels on a prolonged trip into space. In his seminal book, *The Case for Mars*, Robert Zubrin outlined the minimal dangers radiation would pose, placing them in perspective. On a round trip to Mars lasting 180 days (favoring a Conjunction mission over a longer Opposition mission), the average mission radiation dose would be 52 rem (roentgen equivalent man—an estimate on potential health effects of radiation on the human body).

Predicted Radiation Dosage Expected on a Mars Mission (Conjunction results)	
Cosmic rays in transit	31.8 rem
Solar flares in transit	5.5 rem
Cosmic rays on Mars	10.6 rem
Solar flares on Mars	4.1 rem
Total average dose	52.0 rem

(source: Zubrin, R. *The Case for Mars*)

Zubrin calculated that for every 60 rem of received radiation over a prolonged period such as a Mars mission, this would add a one-percent chance of extra risk of fatal cancer to a 35-year-old woman in later life. For a 35-year-old man, the extra cancer risk would be from 80 rem. So, a round trip to

Mars would not necessarily be the radiation monster it has been made out to be—and less so for those staying.

But on the journey to and from Mars, how can Starship better mitigate the effects of radiation to its passengers and crew? Starship will have to protect them from both cosmic rays and solar flares. The former is a "constant but thin rain of radiation" pervading the universe while solar flares are "floods of protons" bursting from the Sun with unpredictability. These solar storms can wreak havoc with satellites and electronic systems on Earth, but cause little harm to humans. However, in space, the Sun's outbursts could shower an unprotected astronaut with hundreds of rem in a few hours, causing radiation sickness and even death.

By the nature of cosmic rays, they cannot be shielded against. However, humans have evolved to live with a certain amount of natural background radiation in the millirem (thousandth of a rem) region, so much so that without enough radiation, a condition called 'hormesis' can develop. So, some small level of radiation is required for our bodies to correctly operate. That said, too much radiation can kill, and thus, astronauts will be exposed to nonlife-threatening cosmic rays on a mission to Mars.

However, solar flares can be blocked out to a high degree to prevent serious radiation exposure. Starship can have installed solar storm shelters, physical barriers near the center of the ship, for maximum protection. Solar rays would then have to penetrate the ship's hull and intervening equipment to get to the shelter, reducing its potency. In some quarters, it was planned that water stores around the shelter would also be an effective barrier to solar flares. The water may be for human usage in addition to 'gray water' solutions and not for fuel production on Mars. But as the "living-off-the-land" approach on Mars calls for minimal resources being brought from Earth, extra "antisolar storm" water may be superfluous. With Starship's large payload area, it would be more beneficial to have the ship's physicality do the work.

Weightlessness

The next biggest issue for the astronauts would be the health hazards, like bone density and muscle deterioration, of prolonged weightlessness in space and even on Mars' low gravity, which is about a third of Earth's.

When Falcon Heavy and Dragon were considered as the Mars mission spacecraft, it had been proposed that during the voyage, artificial gravity would be generated by tethering the Dragon habitat and the TMI (Trans-Mars Injection) booster stage and rotating them about a common axis. This would then generate an effective 1-g of operating environment for the astronauts, saving them from the worst debilitating effects of long-term weightlessness. Whether Starship will have some tethering capability for gravity has not been detailed, yet. However, with the extra room on board, there will be plenty of space for exercise machines and a more expansive medical center. Perhaps a tanker, spent or otherwise, would accompany Starship to Mars and provide the tether craft required.

On Mars' surface, exercise regimes and purpose-built equipment, like exoskeleton suits, would help the astronauts maintain some form of physical strength over time. However, with births among the colonists, humans born on Mars will not be able to visit Earth without extra physical support or other medical advances. Thus, a new branch of humanity would have been produced; the goal of a multiplanetary civilization.

Refueling

As Starship is a reusable spacecraft, like missions to the Moon, it would launch into low-Earth orbit and be refueled by a tanker before heading to Mars. And once on the surface of Mars, Starship will be refueled again for the return to Earth.

One plan had been to launch two robotic Starship cargo flights; the first ship was to be called, "Heart of Gold". The original launch date had been sometime in 2022, with the mission to deliver a large solar panel array, mining equipment, surface vehicles, food, and a life-support infrastructure. And following the Mars Direct plan, in 2024, two years after the first mission, four more Starships would have launched: two more cargo ships and two crewed flights. Their mission would have been to set up the propellant production plant, install the solar panel arrays, construct landing pads, and assemble greenhouses, which could be small, transparent domes.

In regards to the propellant plant, SpaceX intends to adapt the Mars Direct plan. Via their in-situ resource utilization (ISRU) and the Sabatier process, the plant will locally produce methane and liquid oxygen for propellant usage from Mars' atmospheric carbon dioxide and accessible, subsurface water ice. After landing on Mars, Starship would be refueled with locally-produced propellants to return to Earth.

The Systemic Vision for Mars

However, since the 67th annual meeting of the International Astronautical Congress on September 27, 2016, Musk has outlined a "larger systemic vision" for Mars colonization. For Mars to survive as a permanent settlement, there will have to be a complete change of mindset going farther than boots-on-the-ground, flag-planting missions. For a successful, sustainable Mars civilization to grow, the combined actions of individuals, companies, and governments will be required to accelerate and nurture transport to and habitation of Mars. Elon Musk sees the potential of low-cost space transport as being the key to unlocking innovation in other crucial space industries, allowing for further manned missions to other parts of the solar system.

Also, in 2016, SpaceX announced that before Starship's early missions to Mars, prior missions will collect vital data to "refine the design, and better select landing locations based on the availability of extraterrestrial resources such as water and building materials." However, there would be no need for SpaceX to invent new missions for this when current-orbiting NASA and ESA satellites and rovers on the ground could collate and deliver up-to-date information for them. This would encourage Musk's systemic and cooperative vision for Mars missions.

The way things stand, Mars could become a company planet with SpaceX as the primary supplier of the workforce, 'Marsteaders', and tourists. While NASA is focused on the Moon and other commercial space agencies are struggling to get past suborbital flight, SpaceX has a whole Mars settlement concept. SpaceX may be the company concentrating its resources on the low-cost transportation to Mars, but its part in the overall Mars colonization project could see the Red Planet become the Boca Chica of the solar system. Sure, as Musk states, that while every detail for the colonization of Mars cannot be planned, he suspects "a complex adaptive system" will enable others to be independent and choose how they would interact with the growing Mars settlement. But while they may be independent on Mars, until they are truly self-sustaining, they will still depend upon SpaceX for supplies and more people for the near future.

In-Flight Entertainment

You're on the flight of a lifetime to Mars, but what exactly will you be doing for six to eight months? There won't be any Tesla cryo-sleep pods, so a lot of time will be spent in physical exercise; a few hours per day trying to maintain a certain bone density even if there is artificial gravity. It would also be advantageous to maintain Earth sleeping and waking cycles, but perhaps alter it slightly to match Mars' extra 40 or so minutes per day when nearer to the Red Planet. There'll be mealtimes to provide nutrition and create bonding opportunities and briefings either from Earth or the crew regarding flight situations, landing zone conditions, and training on various issues.

Most likely, everything will be in group sessions to ensure a cohesive comradery.

News from Earth, either by email or video link, will be vital for morale, especially the further Starship travels as signals will take longer to arrive the closer Mars gets. Entertainment will be an important boredom and stress reliever and will probably be virtual with games, movies, conversations, and training scenarios taking place in VR (virtual reality), AR (augmented reality), or the metaverse, where joint gatherings can occur in various spaces. Passengers and crew will take part in experiments, or, for academic and entertainment purposes, broadcast updates, news, and blogs/vlogs back to Earth.

There will be alone time, but isolation will not be viewed favorably. On Starship, it will be "one for all, and all for one". And by the time Starship reaches Mars, both crew and passengers will be mentally and physically healthy enough to begin a new adventure for humanity.

Landing on Mars

At a certain point before it reaches Mars, Starship will discard its tether and cease spinning, allowing the ship to prepare for landing. Zero g will return, and the passengers will have to endure the weightlessness again. This may be for a few hours or a day in which to adjust from 1 g, to zero g, and then feel Mars' pull at o.37 g. Then it would be time to strap in for descent.

According to SpaceX calculations, Starship will enter Mars' atmosphere at 7.5 kilometers per second. Its shape and fins will assist in decelerating aerodynamically. The heat shield will defend Starship from Mars' atmosphere, though it may experience some ablation. And then, as per its previous flights, it will orient itself and use the Raptors to descend, touching down softly and vertically. Starship has arrived on Mars.

So, when can we expect the first Starship voyage to Mars? The first option was for SpaceX to send the first unmanned

Starship to Mars in 2024. In December 2020, Elon Musk was "highly confident" SpaceX will land humans on Mars by 2024. A year later, Musk updated that to "Best case is about five years—worst case ten years". Then in March 2022, Musk has now cautiously rolled back the first humans-on-Mars date to 2029. It's still a slide of five years, but with Elon Musk, his Midas touch on technological innovations will see humans on Mars sooner rather than later. And after landing on the Red Planet, how will the first colonists survive?

Chapter 6:

Surviving Mars

"A permanent base on Mars would have a number of advantages beyond being a bonanza for planetary science and geology. If, as some evidence suggests, exotic micro-organisms have arisen independently of terrestrial life, studying them could revolutionize biology, medicine and biotechnology. "

-Paul Davies

Starship has touched down and Mars has welcomed humanity. But there will probably not be a total disembarkment of Starship until the crew and engineers have checked out the remote work of the propellant plant and ensured its operation and safety and also assessed the integrity of the delivered supplies by the cargo ship. Once they are secured, there will then probably be a phased process with people designated jobs and areas of responsibility.

One crew function will be checking the exterior and modules of Starship for any damage from the flight and/or landing. Starship will be home for a while until the base is set up, so it will have to be intact for security and protection. Once the initial settlement inspections and safeguards have been examined and observed, then the real work for survival on Mars will begin.

Colony Settlement

The Mars metropolis may one day develop into a full-scale city from the million or so people who travel there after engineers, scientists, and researchers. It will be one of the greatest endeavors of humankind to start anew on another world, and it will cost a lot of money.

In 2019, Musk anticipated that building a self-sustaining city on Mars will require about one million tons of cargo, brought by the planned 1,000 Starships, at a possible cost of $100,000 per ton. That would equate to a Mars city costing around $100 billion on the low end and up to "$10 trillion on the high end," Musk predicts.

And it may not be just one city. One million people on Mars may be too cramped in a single area, so why not spread out and establish more than one city around the crimson globe?

This was the idea as Paul Wooster, SpaceX's principal Mars development engineer, stated at the 21st Annual International Mars Society Convention in August 2022: "The idea would be to expand out, start off not just with an outpost, but grow into a larger base, not just like there are in Antarctica, but really a village, a town, growing into a city and then multiple cities on Mars." That would be a compelling success for SpaceX to pull off.

For a stable and sustainable base, Starship's landing location should be less than 40° latitude where the temperature is warmer, by Mars standards. This would offer opportunities for best solar power processes and for discovering and tapping into substantial, subsurface water-ice deposits. Other base location considerations would be:

- Topography of the area for settlement expansion, sight lines, and transport mobility.
- Other resource availability, such as minerals and building elements.
- Scientific-rich research areas searching for life, geological investigations, or other endeavors, which would last for years.
- Closeness of other bases to one another so colonists are not overlapping exploration efforts and overexploiting resources.

SpaceX had early design concepts for "green living space" habitats complete with carbon-fiber-frame, geodesic domes, glass panes, and a small force of miner/tunneling droids for the construction of underground, pressurized space for manufacturing operations. The colony will also have to regulate power, lighting, communications, heating/cooling, water purification, air filtration, dust prevention, sewage processing, and oxygen recycling. It may be a good idea to have some manual control over these life-support systems rather than leave everything to a computer, which could break down.

Shelter

The first habitats on Mars will be their own Starships with viable life-support systems. But once the construction crews have built the first base, then it will be time to venture out. After Starship, there will be four types of habitats: rigid, inflatable, tensile, and natural.

Rigid stand-alone Habitat Features:

- Solid aluminum walls, domes, decks, air locks, etc.
- Expensive.
- Difficult to transport and relocate.

Inflatable (e.g. pressurized, domed tent) Habitat Features:

- Polypropylene (common plastic) fabric reinforced by hard frames, such as kevlar.
- More lightweight, portable, and cheaper.
- Susceptible to internal gas leaks, leading to collapse of habitat.
- Susceptible to damage from micrometeorites.

Tensile Habitat Features:

- Similar to inflatables but with internal rigid support.
- Moderate cost and extra handling of support framework.
- Framework supports habitat, even with gas leakage.

Natural Habitat Features:

- Lava tubes/volcanic caves formed after lava has emptied underground tunnels.
- Lower Mars gravity would enable larger cave formations.
- Offer good, natural, structural integrity and protection from changing temperatures and radiation.
- Can be sealed and pressurized with physical sheets or spray applications.
- Can be labor-intensive to make them habitable.
- May not be located in easily-accessible or convenient places.
- Rare chance may be sharing caves with extinct or current Martian life, which humans may contaminate.

Underground/Cut and Fill:

- Using bricks or cement, arched vaults can be constructed and then buried to stabilize against pressure differences.
- Form protective, underground shelters against dust storms, radiation, and temperature fluctuations.

- Large pits can be dug into regolith with prefab modules lowered in and covered.
- Pressurized 'cut-and-fill' habs with stepped or ramped entrances.

Making shelter materials like bricks rather than living in prefab habitats will be another option. You can take Martian regolith, grind, water, and compress it into a mold, and bake it at around 300 °C or even let it freeze into a permafrost block (for use in unheated buildings). For cement, you can use Martian-sourced gypsum, bake it to create lime, and mix with regolith.

However, brick constructions must be kept under compression to prevent the internal air pressure of the structure, which is needed for human survival, from exploding in the lesser, external pressure of Mars. So, it would be a good idea to cover the brick structure in more regolith to compress and maintain the structural integrity.

Radiation Protection

We had noted radiation levels from solar flares and cosmic rays which could present issues for Starship, but what about on the Martian surface?

Solar flares

Mars' thin atmosphere has enough carbon dioxide mass in it to shield the surface from the worst of the solar flares. But even if the Martians are wary of solar flares, staying inside would provide extra protection.

Cosmic rays

These present a bit more risk, but with the planet protecting you from below and the atmosphere offering some protection from above, it is reckoned that cosmic rays will present a 10 rem dosage rate per year, if you are unshielded. For extra protection, you can surround and cover your hab with half a meter of regolith sandbags or boron-infused (borated) water ice, which will reduce the radiation dosage rate to six rems per year.

Power

The pioneers from Starship will also have to establish energy centers from which to power settlement processes and permit manufacturing of base and personal items. Solar, nuclear, and wind will be the primary sources, though fossil fuels may be used to drive terraforming efforts.

Solar

Starship could deliver the parts for ground-based solar panel arrays to produce sufficient power for a small base, whether it be a centralized solar farm or a localized network of panels.

However, solar panels can be covered in dust for weeks on end due to Mars' dust storms. In December 2021, a dust storm struck the United Arab Emirates-led Hope mission, spreading over its location within the thousands-of-miles-wide crater and also affecting NASA's Perseverance rover and Ingenuity helicopter, as well. The storm lasted until January 14, 2022. Knowing this, a base could not rely completely on solar power.

Regular cleaning and maintenance of solar panels and the monitoring of dust storms patterns would have to be a primary job. Plus, there is less light from the sun at Mars' distance, so panels would have to be huge and spread out over larger distances for the colony to achieve reliable power. And there would be no power at night, unless energy was stored. Thus, solar panels should be used for smaller installations (possibly unmanned) or as a storage backup.

Nuclear

Alternatively, NASA has been researching fission reactors for use in deep space missions. For a Mars base, it is estimated that for sufficient, fission-reactor-based electric power, its mass would be between 210 and 216 metric tons (463,000 and 476,000 lb), requiring at least two Starships journeys.

However, compact nuclear plants, as used on Earth ships and submarines, would also suffice until larger plants could be constructed, with enough uranium transported from Earth. The nuclear plants would provide power from its heat generation either as heat or converted to electric power. The plant would have to be protected by heavy-duty containers and any leaks contained, though on Mars, they would likely be localized. However, though nuclear power would be a better source of energy for areas where wind and solar generation is not ideal, nuclear power plants would also require consideration for storage of spent fuel and decommissioned reactors.

The more-likely and safer nuclear option would be for small radioisotope units to be used around the base, buried a few hundred meters away (and away from underground water sources) to avoid contamination should there be a containment leak. These would power the propellant plant and other necessary services.

Wind

The winds on Mars are quite strong, but in a third of gravity compared to Earth, their force is lessened. As with solar power and the Sun's strength, turbines would rely on the wind blowing, but stored energy may mitigate this shortcoming.

But a wind power system using a vertical-axis turbine to produce sufficient power might amass around three tons (6,900 lb), enough for Starship to deliver multiple units.

Geothermal

There may still exist possible heat activity close to the surface from Mars' volcanic past. It may require time to prospect for such active points, but once found, the site can be drilled and pumped with water to produce steam, which can be extracted to power turbines. But this process will be labor-intensive with infrastructure set up, such as the drilling equipment, pipelines, and cables for power conveyance.

Natural Gas

Methane is located around three equatorial locations leaking into the atmosphere, though the source is a mystery as to whether it is natural or geological. But even more mysterious is the fact that methane suddenly disappears by some unknown process. If the methane source is natural, then life will have been discovered on Mars. Equally, there may potentially be reservoirs with millions of tons of nonbiological methane, which the colonists may be able to tap into. This would enable the methane to be used immediately as rocket fuel, power rovers and land vehicles, and used in other chemical processes, such as to produce plastics. Either way, methane will have a profound effect on the Martian colonists' lives.

Life Support

Mars' colonists could decide between or split the use of two types of life-support systems: biological/bioregenerative or physical-chemical. These systems will help provide oxygen and clean water, remove CO2 and waste, and in some cases, provide food.

Biological

This uses natural agencies like animals, bacteria, and plants to generate air and recycle human waste, plus provide food. But it is a complex, unpredictable system relying on networks of living organisms to compost, purify water, and defy behavioral models. Managing biological systems requires constant upkeep, so it may be too laborious for a small colony.

Chemical

The application of chemical reactors can be used to split oxygen from carbon dioxide, creating air and also heat to burn human waste. It is a more predictable system, but more power-hungry to run systems such as oxidation, pyrolysis (heat incineration of waste), and decomposition processes. Despite this, the colonists may choose this method for its reliability.

Resources such as water and air can also be derived from other sources, both planetwide and in localized areas. Extraction techniques used on Earth would also be ideal here.

Water

The most abundant water source on Mars for colonists, away from the poles, will be in the permafrost with water ice mixed in with the regolith. This can be mined and the water extracted, but it is a lengthy process and probably only good for localized water collection.

Ice deposits at the poles would hold a greater volume of water, but its distance from more equatorial settlements may make collection problematic, unless the base is located there. The ice could be kilometers deep and will be as hard as granite, so specialized drilling equipment will be required.

Subsurface aquifers may still be active from Mars' heat and pressure. But as with deep-lake drilling in Antarctica, the aquifers will have to be investigated for signs of life beforehand to avoid contamination. Aquifers may only be located in a few locations, so finding them will be a tremendous endeavor with infrastructure built up around it to transport the water. Once extracted, the water will require purification before usage.

Another common source of water will be the atmosphere itself. But the extraction efforts will be an arduous process, with Mars' dry atmosphere yielding a kilogram of water per one million cubic meters of air.

Air

Mars' atmosphere is 95% CO2, 3% nitrogen, 1.6% argon, with trace amounts of oxygen, water vapor, and methane. To extract oxygen, electricity can be used for the electrolysis process to free oxygen from H2O. The hydrogen can then be used for fuel processes. Oxygen can also be removed from atmospheric CO2.

Food

Growing their own food may suit colonists as more palatable than crops grown using recycled human waste as fertilizer. Greenhouses will be able to be localized with crops grown under artificial light depending on how much expansive power can be used for this. Sunlight on Mars may be 40% lower than on Earth but still available for greenhouse purposes.

Colonists will have to balance the internal air pressure and also anchor the greenhouse to the ground to prevent it from collapsing. It will also allow for both proper plant growth and for unsuited work, plus the higher pressure would allow insects to be introduced to pollinate the plants rather than to do so by hand. Crops will need the right range of pressure, nitrogen, oxygen, water vapor, carbon dioxide, other trace elements, and soil to grow. Typical crops that can be grown on Mars would be fruit and vegetables, but to use the whole of the plant efficiently (including roots, leaves, stalks, etc.), it may be good to have livestock on Mars to eat those parts, like chickens and omnivorous goats, which would make good protein sources, as well.

Hydroponic systems can be applied to grow plants without the need for soil. The roots of the plants are directly fed nutrients dissolved in water. The roots can be suspended in water or within sand or gravel for the nutrients to circulate through them. There's no need for fertilizers, water is used sparingly, and plants can be grown closer together. Hydroponic buildings or rooms, whether on the surface or underground, can provide artificial light for the plants in a warm, humid environment.

But can crops be grown directly in Mars' regolith? Colonists will find that Mars' soil is devoid of biological material to make the ground fertile, unlike on Earth. While some crops may grow in Martian soil, some test samples of soil by 1970-80s Viking lander missions showed that some soil regions would be toxic to plants, whereas the 2008's Phoenix lander mission revealed soil that was alkaline enough to grow certain Earth crops. However, botanists would be able to test and chemically treat the soil for plant growth or add organic products for the same effect.

Meat

Apart from taking a risk on goats and chickens, meat protein on Mars may also come from fish, crabs, shellfish, and shrimp. Insects and snails may also be cultivated for Martian delicacies. But perhaps Mars may be the first place to routinely eat artificial-meat, lab-grown protein which can be molded into recognized foods like burgers, sausages, and steaks. 3D-printed meat techniques are also on the rise. They may be pricey on Earth menus, but Martian chefs may employ this method to entice people to eat cleverly-disguised snail steaks.

Travel & Exploration

Ground travel

Cybertruck

Lest we forget, SpaceX has already sent a 'passenger' toward Mars. On February 6, 2018, Elon Musk's Tesla Roadster was launched into space as the dummy payload for the Falcon Heavy test flight. A space-suited mannequin, dubbed 'Starman', was the sole occupant of the car. Nine months after launch, the Tesla had traveled beyond the orbit of Mars, its closest approach being eight million kilometers (five million miles) away in October 2020. Unfortunately, Roadsters won't be driving on Mars anytime soon.

But maybe a version of Tesla's Cybertruck will. In November 2019, Elon Musk unveiled the all-electric, shiny-silver, sharped-contoured Cybertruck. It was constructed with a near-impenetrable, ultra-hard, 30X cold-rolled, stainless-steel exoskeleton shell and Tesla armor glass for endurance, durability, and passenger protection. It has a payload capacity up to 3,500 pounds, adjustable air suspension, and 100 cubic feet of exterior, lockable storage. It can seat six comfortably, and as a Tesla product, has an advanced, 17" touchscreen with a customizable user interface. It is the rugged pickup of the future. There had been comments that Cybertruck was Musk's Mars rover, but there were questions over pressurization (which it wouldn't need to be, as discussed below). However, in interviews, Musk has called Cybertruck the "official truck of Mars" and hinted that Starship could include several pressurized versions of Cybertruck.

Rovers and Light Vehicles

The rule of thumb may be to have a light vehicle for Mars, an ATV or motorcycle, over a large, pressurized vehicle, unless you're on a long-haul science mission. Pressurized rovers can vary in size, though most will be slightly larger than an Earth-based SUV. The advantages of a pressurized rover are you can relax within an enclosed environment around you while driving and provide a shelter for when you rest. But as with Earth vehicles, rovers can be heavy, too big for accessing some regions, and could get stuck. Also, unless the pressurized rover is big enough to have an air lock, if you see anything to investigate and require samples from outside, it would involve putting on your space suit, depressurizing the rover, going out, completing the mission, getting back in with added Martian dust on you, repressurising, and off you go. You'd then have to clean and decontaminate the rover upon return to base.

With unpressurized vehicles, you'd have to be in your space suit and relying on a smaller oxygen supply. The rover can be smaller in size, but similar to a moon rover and more robust in Mars' higher gravity. But they would be more mobile and lightweight than a pressurized rover. If you get stuck, you may be able to push/pull or lift the rover, ATV, Mars cycle, etc., out of danger. Also, if you saw anything to investigate and required samples, all you'd have to do is hop off the rover or bike, collect it, put it in a storage container, and move on. Plus, being in an unpressurized vehicle would bring you closer to the Martian environment. For shelter with an unpressurized rover/bike, you can tow a trailer with supplies, like an inflatable tent or bag with rations, or stow them in a storage bin.

Your power source for a rover, bike, ATV, etc., will be chemical rather than nuclear or solar. A radioisotope power source would offer almost limitless traveling distance, but will require shielding, thus more bulk for the vehicle and leak issues. Solar power on a dusty planet might not be so good, as the panels could be damaged, not work at night, and/or be covered with dust. It would be better to opt for a mutually-compatible power source that runs both the space suit and the vehicle: methanol/liquid oxygen fuel cells. These cells can provide your life support, run the vehicle, and in emergencies, the vehicle's fuel tank could provide oxygen and water for your survival.

So, for expediency and the chance to explore your own way, a one or two-person, unpressurized, chemical-powered vehicle would be the way to go on Mars. No doubt, Tesla and SpaceX would provide a suitable, land-based, vehicular power source, one that may also pollute in a beneficial way to terraform Mars.

Air travel

While Ingenuity, the little helicopter on Mars carried by Perseverance on February 18, 2021, has wowed Earth audiences with its record-breaking flights, something more substantial will be needed for humans.

Balloons may be a good way to get around on Mars, with the atmosphere being more naturally suited to balloons, as the Red Planet lacks the complex air circulation Earth does. The balloon bag size will be bigger to compensate for Martian atmospheric density, plus you would have to choose the best gas to use for your bag, whether hydrogen, water vapor, or heated CO2.

Dirigibles with streamlined shapes, lifting gas (e.g. hydrogen), and engines for lift and steering would offer a better transport option than balloons, providing a greater payload over a longer distance. Airships could also be installed with solar cells on the upper surface to power an electrical drive system. Though traveling may be slow, there wouldn't be a need to refuel to look for a long landing strip.

Airplanes may take flight on Mars with people opting for subsonic (c 700km per hour) or supersonic flight. The large, long, straight wings of a subsonic plane can allow for superior lift-to-drag ratio even using propellers, but long landing strips may be less frequent. So, the use of tilt-rotors or ventral jets from rocket engines for vertical takeoff on a supersonic plane may be more advantageous.

But if you do not have time for a slow journey or just want the thrill, then a rocket-powered aircraft would be the best option.

Like a missile, the aircraft would launch steeply into the air and trace a computer-guided, parabolic path to its destination, landing with the aid of parachutes and thrusters. Supersonic (faster than speed of sound) and even hypersonic aircraft (five times faster than speed of sound) may also be available, but why miss the grandeur of the Martian landscape?

Tesla Bot on Mars

Another Tesla product which may one day grace Mars is Tesla Bot or Optimus. Elon Musk unveiled the Tesla Bot concept on August 19, 2021. For a more detailed explanation of Tesla Bot, check out the book, Tesla Bot Optimus: Elon Musk's Robot and the Future of Work.

Because the Tesla Bot requires no oxygen or food, it could potentially play a pivotal role in colonizing Mars. Making humans a multiplanetary species has been a decades long goal of Elon Musk, and there are many jobs to be done if this were to become a reality. Optimus would be ideally suited for the Mars environment since the robot not only requires no oxygen, water or food but it can also tolerate a broad range of climate fluctuations while never getting sick.

Optimus is the size of a small human at 5 feet 8 inches tall and weighing 125 lbs It's able to lift 150 lbs, and walk at a speed of five miles per hour.

The Tesla Bot is designed to be an autonomous humanoid robot (AHR) tasked to perform general robotic work for humans and complete dangerous, repetitive, and boring tasks. This nonthreatening "buddy robot" would also be a future companion for someone who needs to feel less lonely or need help completing work, like someone on Mars.

On Mars we are likely to see Optimus on construction sites, in power plants, digging tunnels, and carrying out other maintenance work. Monotonous construction work such as scaffolding erection, brick and tile laying, plastering, cement laying, carpentry, etc. could be outsourced to humanoid robots. Though additive manufacturing is gaining ground in constructing viable housing and structures, Tesla Bot could assist with this, or build a dome on Mars or house to specifications from plans designed by an architect.

Another advantage of Optimus is that as a humanoid robot rather than a specialized one, it would be able to use the same tools as humans instead of having customized tools made for it. Tools become dual-use for humans, and AHRs bring down tool costs and reduce storage space. At the first opportunity, we'll need to send Tesla Bots to help with all of the work to be completed on Mars.

Power Plants require constant attention, and engineers are not always on-site or able to attend immediately in case of an emergency. While most modern sites have smart systems such as Building Management Systems (BMS), it would still be prudent to have a Tesla Bot on-site to monitor the plant and equipment and deal with any changes or emergencies. This would include hazardous sites like nuclear plants where robots may be able to enter contaminated areas (depending on its remit, discussed further in robot job law and legislation).

The type of tasks undertaken by Optimus would be general manual labor tasks: loading and lifting. Courier roles and errands delivering materials around Mars would be ideal for autonomous humanoid robots. Robots could benefit from not requiring space suits, oxygen and being able to withstand radiation.

Although the role of Optimus was mostly delineated along the lines of completing repetitive and boring tasks, this rather wide vision can be expanded into various roles in the domestic setting. Cleaning (domestic and industrial areas) would be an obvious role, whether it be windows, streets, construction areas, floors, factories or houses. We do have vacuum bots wheeling around the house hoovering up dust and bumping rather amusingly into furniture, pets, and people. A humanoid robot could be more advantageous in reaching higher spots and providing a thorough clean without getting bored or distracted, always giving one hundred percent, and never conveniently forgetting to clean the less appealing areas of the apartment or factory. And Optimus would also need to be trained in chemical analysis or have attached sensors or cameras able to discern contaminated areas (to detect different scents, for example) and apply the correct cleaner or solvent.

Optimus could also care for humanity on Mars in other ways. Optimus would be unaffected by viruses, at least human-made or biological viruses, not malware. Imagine factories and companies able to keep running during pandemics because they have a Tesla Bot workforce, which can be washed down with anti-viral solutions, not be affected by lockdowns and furloughs, and offer services to keep the economy moving along while humans recover from any sickness or pandemic on Mars. They would require minimal or remote human supervision. Optimus could not only coronavirus-proof humanity from food and supply shortages, with robots continuing to work during lockdowns, but also assist during sand storms, asteroid collisions or natural disasters.

Light maintenance work could be carried out, such as fixing walls, checking pipes for leaks, landscaping, or building tunnels or homes. Repetitive, error-free tasks would ensure quality control on projects, and lower maintenance costs.

Over the years, Tesla has developed an impressive array of technologies for their cars including Full-Self Driving (FSD) computers, the DOJO supercomputer which trains its deep neural networks for artificial intelligence (AI) advancement, batteries that are becoming more powerful and energy-efficient, and other sensors, cameras, and equipment to enable their cars to drive fully independently. It is these impressive technologies that Musk will incorporate into the Tesla Bot on Mars, ensuring a successful transition from Earth to Mars.

In the past, there had been many different types of industrial robots, but they were all specialized for a particular task, such as welding or painting. Before Tesla Bot there was no robot that could do the range of tasks that a human worker could do.

The robot's energy requirements are significant as it needs to be able to power itself for long periods of time. Solar and nuclear power on mars would charge the Tesla Bot's batteries.

With their cutting-edge battery technology and other innovations leading the way on EVs, making future Tesla cars three to four times more efficient than internal combustion engines, Optimus will be installed with an upgraded version of this battery pack. The battery would be lightweight, durable, fire-resistant, exchangeable if required, easy to charge, easy to troubleshoot, and eco-friendly.

Not resting on its laurels, Tesla still conducts battery research to enhance its capabilities and to understand its shelf-life. For Optimus to be efficient and long-lasting, the battery will also have to perform. Tesla's investment in lithium-ion battery research consists of research and development partnerships and developing next-gen batteries. Tesla wants cheaper, smaller, and lighter batteries for its vehicles with new designs and resources, which ultimately will pay dividends for the performance of Optimus on Mars.

When Tesla first developed their cars, they installed an advanced driver-assistance system called Autopilot, initially in 2014, and upgraded in 2016. It is a sensor system combining both hardware and software which led to an earlier form of full self-driving support. The system consisted of eight cameras, twelve ultrasonic sensors, and forward-facing radar. Each subsequent upgrade provided more powerful computers with a custom Tesla-designed chip system. In 2019, Tesla was ready to announce that as a standard feature all of its cars would include their Autopilot software. The radar system was soon removed so the cars relied solely on their cameras. Musk's reasoning for this was that "humans could drive with only two eyes and that this meant cars should be able to drive with cameras alone." Such a logical deduction, though not entirely without shortcomings, could be the basis for sight processing for Optimus.

Following on from Autopilot came its successor, the full self-driving system which would facilitate fully autonomous driving. Testing began in 2016, but as of July 2021 full autonomy has not been successfully demonstrated. This is because Tesla is completely relying on their camera system and not radar or lidar to navigate, a strategy that some experts deem unfeasible. The camera system is not as detailed when compared to other companies such as Waymo or Cruise which are using much more detailed three-dimensional maps, lidar (Light Detection and Ranging — a remote sensing method to make digital 3-D representations by targeting an object or a surface with a laser and measuring the time for the reflected light to return to the receiver), radar, ultrasonic sensors, and cameras for their autonomous vehicles. But Tesla has persevered with drive-testing their self-driving software for over 20 billion miles as of January 2021.

To ensure an autonomous driving experience, Tesla has designed and installed a full self-driving computer chip in its cars since March 2019. But, as of yet, there has been no successful demonstration of the FSD system, prompting the National Transportation Safety Board (NTSB) to call for "tougher requirements" placed on any autopilot testing on public roads, and for Tesla's Autopilot system to be redesigned. However, while the system is not yet capable of autonomous driving, Tesla will continue their redesign, testing, and perfecting. Currently, Tesla admits they are only at Level 2 vehicle automation whereas Level 5 is full autonomous driving.

How the FSD computer system will figure operationally with Optimus if it cannot work with vehicles has not been fully explained. Of course, Optimus will be a bipedal product, different from a wheeled entity, so adjustments will have to be incorporated into the design, application, and testing. In the current Tesla Bot model design, the FSD computer will comprise a large part of the torso with multiple autopilot cameras to be installed for sight and navigation. Will Tesla have to backtrack and resort to installing radar, lidar, and other sensors to ensure a fully autonomous Optimus or will they persist with a camera-only system connected to neural networks?

Tesla's neural network system is still cutting-edge research applied "to train deep neural networks on problems ranging from perception to control." This implies that Optimus will learn how to perceive and study high-quality images in order to detect objects, determine details, and discern depth in a 3D worldview. With Autopilot algorithms also creating planning and strategy systems for the neural networks, Optimus will learn a diverse array of situational events so it can navigate and react safely and accordingly. For a Tesla vehicle, they have stated that a "full build of Autopilot neural networks involves 48 networks that take 70,000 GPU hours to train. Together, they output 1,000 distinct tensors (predictions) at each timestep." Evaluating this for Optimus, which will arguably be in a denser object environment, competing for space with humans, the neural network system will undoubtedly have a richer training process. Optimus should be able to evaluate and utilize data from its learned neural network system in order to operate in complicated real-world situations.

The FSD and the neural network system are not the only systems charged with educating the Tesla Bot about its environment. Tesla has also developed DOJO.The Tesla D1 chips, designed and manufactured by Tesla, will have over an exaflop (a million teraflops) of processing power, which Tesla claims will make Dojo the fastest AI-training computer as compared to competitors like Intel and Nvidia. Dojo has been designed to make Tesla AI learning more efficient, training the neural networks on huge amounts of data, as mentioned above, so the more data the AI is trained on, the more efficient and better it becomes while the AI learns from its mistakes. For a vehicle testing in self-driving mode, it will have the data transferred from its learning environment back to Dojo. Dojo would then use the data to train the neural network in real-time.

With an in-house supercomputer, Tesla can train the networks for both their autonomous vehicles and robots. Optimus will learn how to walk, navigate, and perform tasks from data gathered through Dojo. This could give the Tesla Bot an advantage, as Tesla has had years of deep network learning through Autopilot and the FSD, which Dojo will accelerate. In the future, there will be a mini-Dojo inside each Tesla Bot. Dojo's brains are working on it full-time.

The robot has an outer shell made of flexible soft skin that helps with dexterity and sensors. Underneath that skin are 28 joints throughout the robot's arms and legs. These will give Optimus a full range of motion and enable the robot to accomplish most tasks that humans perform.

There was also an interesting choice with the face in that there will be no human-looking face, no flashing eyes, or cameras embedded in the head. Instead a screen will be used on the face of the robot to display useful information. The Autopilot cameras will be in the head and hidden within the face screen. The current color scheme of the model may not be the final iteration, but a white and silver body with black shoulders, neck, and head will make Optimus stand out. Most likely, color schemes may change according to different functions or jobs on Mars.

The face is made from Tesla glass technology which they use on the roof glass for the Tesla Model 3. The glass technology also produces the glass used in the Tesla Solar Roof solar panels. So, if installed on the face screen or other parts of the Tesla Bot, there may be solar panel capabilities. The face panel could also double as a touch screen with or without one on the chest. Having a face touchscreen would introduce possible communication possibilities such as a phone function, GPS, App features, and data input/output ports. Optimus, as well as being a friendly human helper, could also double as an autonomous communications and data center.

The skeleton of the Tesla Bot will be of lightweight materials, which will incorporate the FSD computer and actuators. The material of the physical structure may be a composite of new materials. The Tesla Roadster was made of carbon fiber, while the Tesla Model S was made of lightweight aluminum. Tesla has experimented with material technology so it may not surprise anyone if they created a new material for Optimus. But whatever the material used, in line with Musk's ethos for sustainability, Optimus' material supply chain will have to be ethically and sustainably sourced, mined, produced, and recycled.

The Tesla Bot will have human-level hands, and two axis feet for balancing with force feedback sensing. Moving all of these components will be 40 electromechanical actuators. The arms, legs, and hands will each have 12 actuators, while the torso and neck will each have 2 actuators.

Tesla has already made significant advancements in the development of artificial intelligence (AI), autonomous driving, and machine learning—the three core technologies that will enable the robot to carry out human-like tasks.

Both crewed and cargo Starships could carry a complement of Optimus robots stowed away until Mars is reached. There, Optimus could help set up communications, power, exploratory equipment, and base infrastructure before humans arrived. They could also carry out short, planet-side, exploratory missions into hazardous areas. But whether a bipedal robot could safely traverse Martian terrain may be another matter, with NASA and other space agencies preferring tracked, wheeled, or quadruped robots over a humanoid robot for long-range exploring. Optimus would be better suited for the on-site building of the base, cleaning, maintenance, etc.

Musk has said that instead of Tesla being thought of as a car manufacturer, it should actually be considered the world's largest robotics company. After he said this and added that the robot project will grow to become Tesla's most important product ever produced, most people didn't take his claims seriously. He then repeated these predictions on a quarterly conference call with Wall Street analysts. He stated on the call that too few people yet realized the importance or implications of this new robot. He said that he is confident that Optimus will grow to not only be Tesla's most important product ever produced, but will also completely transform the labor economics of the global economy. It will make products cheap and abundant, and usher in a world of new abundance. Musk then repeated the claim that he was confident that Tesla's new Optimus robot will generate more revenue for the company than the sales of all of their electric cars combined. And if that wasn't enough, he also predicted that robots will one day become the dominant source of global labor and droids will replace humans as the new working class.

This will transform our current understanding of what an economy even means. In a capitalist system, all capital is simply the distillation of labor. When labor becomes cheap and abundant, building structures and colonizing Mars becomes much easier.

In many SciFi stories, the whole intelligent robot thing doesn't work out so well for the humans involved. Therefore, how can Tesla overcome these grim forecasts and keep Optimus from going rogue and enslaving everyone on Mars or harming people? Elon was able to speak candidly about his approach to the Optimus project in a recent interview with Business Insider. He said that, "With regards to AI and robotics, of course I view things with some trepidation because I certainly don't want to have anything that could potentially be harmful to humanity. However humanoid robots are becoming a reality. Consider the case of Boston Dynamics. They improve their demos each year as AI advances at a breakneck speed." Elon states that since artificial intelligence in robots will occur with or without him, then he wishes to lead the charge himself since he lacks confidence in others to do it responsibly.

While the guidelines and code for running a safe robot should be simple enough to establish, the prevention of hacking, misuse, error and ethical issues will be much more difficult to sort out. To help counter these threats, Elon stated that the design of Optimus will be physically unemposing. The obvious goal of this design is that most ordinary individuals should be able to outrun or overpower a Tesla Bot in the event the robot goes rogue and tries to harm them. This is in sharp contrast to a robot-like Atlas, produced by Boston Dynamics. Atlas is a much stronger and faster robot weighing almost 200 pounds and can outrun most people, jump and perform parkour and gymnastics, and would be able to easily harm a person, if it were so inclined.

The Optimus robot is not intended to be spectacular or gigantic. Rather, its design is to be functional and practical to fit into everyday Martian life. It's been referred to as a worker droid by Elon at Tesla's Austin, Texas Gigafactory's Cyber Rodeo event. The Tesla Bot is designed to gradually take over many of the mundane, repetitive and dangerous jobs that make up the backbone of the Mars economy.

Naturally, the ultimate key to a safe and effective robot is not in the design but in the programming. This is where Tesla truly excels. The company's experience with real-world artificial intelligence programming positions Tesla as the front-runner in the race to implement a truly artificial human robot. The same technology used to create driverless cars capable of navigating city streets alongside human drivers will be used to create robots capable of seamlessly integrating into our everyday lives.

For hundreds of years, machines have been replacing human jobs and robots have already begun displacing human workers in several industries. For example, today most of the welding and painting of cars is done by giant robotic arms.

Returning to our favorite works of science fiction, we find Isaac Asimov's book from 1950 called I Robot. The book takes a nuanced look at sentient machines through a series of short stories. Asimov established three guidelines for robot safety. The first rule is to avoid causing injury to a person or allowing harm to come to a human through inaction. The second rule is always to obey a human's command, unless it conflicts with the first rule. The third rule is self-preservation, as long as it doesn't conflict with rules one or two.

Taking a big-picture view, there are only two paths forward for robotic automation on Mars: either attempt to replace every human task with a specialized robot designed for that task, or we replace the human with a robot capable of performing all tasks. Elon effectively said this to Business Insider, stating that humans constructed the world to interact with a bipedal humanoid with two arms and ten fingers. Therefore if you want a robot to be capable of performing all of the tasks that humans can do, then it must have roughly the same size, shape and capability as humans.

Elon Musk said he envisions a future with a Robot in every home and one taking care of all your needs so you can live a life of leisure. But AI robots will not come without risks. "By building artificial intelligence we are summoning the demon," said Musk. There are many people who believe that by building autonomous robots, we risk providing a future super intelligent AI with millions of mechanized soldiers.

Bloggers at The Mars Blueprint crunched some numbers for the inclusion of Optimus on a Starship mission to Mars. If Elon Musk has the foreknowledge in proclaiming Optimus' capabilities and operation by the time of the first Mars missions, then its inclusion would be advantageous.

The Mars Blueprint compared Optimus' size and weight to another SpaceX Mars-bound product—that of a Starlink satellite. With a cargo-loaded Starship, how many Optimus' can be loaded on board? A present-day Starlink satellite weighs about 600 lbs (272 kg) and 60 fit in each Falcon 9 launch, so an estimated 400 satellites would fit in Starship's payload bay. Five Optimus robots weigh slightly more than one Starlink satellite but are around the same size, if stowed away efficiently. So, 20 Starlink satellites would approximately equal 100 Optimus robots. So, if Optimus was a viable AHR ready for work on Mars before or during the human colonists' arrival, then this would be a sizable workforce able to set up the base.

With the help of autonomous robots, you have survived the first few months on Mars, settled into a routine, and feel your life is safe enough to put down roots and relax. The Red Planet has much to offer, and it is time to spread your wings as the community grows. Survival mode has turned into life as usual.

Chapter 7:

Life on Mars

"We've gotta become the Martians. And we've gotta go to Mars and civilize Mars and build a whole civilization on Mars and then move out, 300 years from now into the universe. And when we do that, we have a chance of living forever."

-Ray Bradbury

The more people added to the millions already on Mars will create a situation where more social and physical infrastructure is required to continue with normal life. Just as Earth civilizations grew from simple settlements, so will Mars follow the path of complexity. Even with smaller, independent settlements dotted around, there will have to be institutions to cater to and for its population. In order for there to be less chaos and conflict over rules and regulations, there may have to be a centralized authority, whether it answers to Earth or not. For a self-sustaining human civilization on Earth, the trappings of such bureaucracy will be mirrored on Mars.

50 Year Plan for Mars

"Mars is there, waiting to be reached. I think humans will reach Mars, and I would like to see it happen in my lifetime."

-Buzz Aldrin

Colonizing Mars and making humans a multiplanetary species won't happen all at once. There's a phased approach and a detailed plan to make this dream a reality.

Like all of the other planets surrounding Earth in the solar system, Mars is a wasteland free from all forms of life. But

50 years from now, Mars could be teeming with a new human space colony.

Thanks to its closeness to the sun and its lack of volatile weather compared to other planets like Venus, Mars is seen as a possible site for human colonization.

But how exactly would people transform the red planet of dust and rock into a thriving human city? Although it won't happen overnight, the foundations for this plan are already being laid today.

In 2016, Elon Musk, who had been fascinated by the idea of colonizing another planet for a long time, took his first step towards making this plan a reality. He announced the first designs of the rocket that would deliver people and supplies to the surface of Mars.

He called it "Starship". That summer, his team began working on the technology that would make Starship possible. The same year, the Russian aeronautical company Roscosmos successfully launched a probe into the orbit of Mars.

In 2018 humanity was still a long way from Mars, both literally and metaphorically. There were many obstacles in the way. The surface of Mars is deeply inhospitable to human life. While humans can survive in space suits, there is no food on mars and the journey would be very long and challenging.

This same year NASA took the next step, they sent a robot to Mars called the InSight Lander. This was the latest robotic probe being sent to Mars, but this one had a special mission. It would place an electronic seismometer on the planet to map it more closely. It would look inside of the planet to get a picture of its internal heat transfer. In order to send humans to Mars, we'll need to understand Mars a lot better than we do now and know how best to survive.

SpaceX has been spending the years since Starship was announced making plans and working on building the ships. In 2019, they unveiled the first proof-of-concept for its fleet of rockets called the Starhopper. The prototype for this test rocket made its first test hops. These short flights were simulating the kind of jumps it'll need to make to get to

Mars. Starhopper successfully completes every flight, and SpaceX is soon building full-scale prototypes to send the first crews of people to Mars. That year several government officials started asking, "shouldn't space travel be in the public's hands and not just a private company?" But except for any large-scale investment, it seems likely that any future space exploration will be led by private companies and not governments.

2021 was a huge year for space exploration. The United Arab Emirates (UAE) successfully launched a probe into orbit around Mars. Also, NASA successfully landed their Perseverance rover on Mars to collect samples. They also sent along a small drone-helicopter.

As SpaceX completed a high-altitude test flight and landing of the latest Starship prototype, NASA officially selected the company to build the first human lunar lander model. This greatly angered Elon Musk's biggest competitor, Amazon founder Jeff Bezos. Bezos has his own space company called Blue Origin. Also later that year, China successfully landed its own rover on the red planet. The race to Mars was officially becoming a global competition.

By 2026 even more countries will join the race. Japan and India, both of which have fast-growing space programs, will get closer to the red planet. India is set to launch its Mangalyaan 2 orbiter to collect data, while Japan will send its Martian Moons Explorer to capture samples from the moon of Phobos and return them to Earth over a four-year mission. The Starship prototype is now operating crewed test flights. As the first orbital flight with a crew is a success, Musk and his team get ready for an even bigger leap. However, one thing still stands in their way, the ideal alignment of Earth with Mars. This only happens once every two years, so timing is key. If Mars colonization is going to happen, everything needs to be just right.

In 2028 SpaceX is ready for a flight to the moon. The "Dearmoon" mission will take off from Earth and bring 8 carefully selected crew members on an orbit around the moon. At the same time, an uncrewed SpaceX ship will be getting ready to land on the moon's south pole. But the most significant development is happening elsewhere, as the main structure of the Lunar Gateway is completed. This small space station, a collaboration between American, European, Canadian, and Japanese space programs, will orbit the moon and be a key waystation for larger space missions in the future.

In 2030 The moment is here, finally making the leap to Mars. Starship has been tested, but this is going to be its longest trip yet, and so the first missions to Mars will be unmanned. These two cargo Starships will head to Mars, both loaded with a miniature nuclear power reactor, as well as an atmospheric propellant plant to create oxygen and methane from the Martian environment. Methane is essential for sustaining both human and plant life on the red planet. NASA will also be retrieving samples from Mars to transport back to Earth. This is also the year that SpaceX plans to return humans to the moon to search for frozen water in its polar region.

In 2032 the Lunar Starship was tested successfully. The goal of these lunar missions will be to establish the first human base on the moon. The moon's Artemis Base Camp will allow for the astronauts to stay for longer periods and do research on-site. There are no plans to create permanent living quarters on the moon, but crews will rather be rotating in and out quite frequently. While this work is going on, the results from the ground tests on Mars will come back allowing the SpaceX team to make their final selection for the ideal location of what will become Mars Base Alpha, the very first human outpost on Mars.

In 2034 two unmanned starships arrive on the future site of Mars Base Alpha. A large fleet of rovers, some solar panels, and mining and worker robots will lay the foundation of the base. It'll be one of the biggest construction projects in the history of humanity - and it will all take place over 80 million miles away from the closest humans. This project will take more than a year to complete.

In 2036 the robots are hard at work as landing fields are cleared and the base continues to be built. While these robotic workers are preparing the site, the moon is about to become a lot more crowded. Artemis Base Camp was completed not long ago, and after several short-term missions for construction and testing, it's now ready to welcome its first human residents. Much like the International Space Station (ISS), this base will allow its residents to engage in longer-term study and to witness the effects of space living on humans. This data will become very important when sending humans to Mars.

In 2038 SpaceX has been hard at work searching for the ideal candidates, and now twenty-four of the selected individuals have been chosen. They're about to become the astronaut crew members of the first manned mission to Mars. They'll be divided into two groups of twelve, and sent on a pair of Starships accompanied by cargo ships. While this is mainly a SpaceX mission, it will also be supported by NASA and watched closely by other governments around the world.

The cargo ships are critically important to this mission, because they will be carrying an ISRU system. This machine will be used to collect, process, store, and use materials found on Mars to replace important supplies and propellants that would otherwise need to be taken from Earth. This ISRU system will save a massive amount of cargo space. Solar panels are also being set up around the planet by robots, providing much of the desperately needed power for Mars Base Alpha. SpaceX also will deploy the Starlink satellites around Mars, providing communication and GPS abilities. While SpaceX estimates that the timeline of the trip to Mars will be significantly shorter than the original NASA voyages, it's still a several months long trip. The twenty-four astronauts are in for a very long and sometimes stressful voyage.

In 2040 the brand new 24 residents of Mars Base Alpha are living in a self-contained environment. With an artificial atmosphere, and only traveling outside the base for research missions under carefully controlled circumstances in space suits that can withstand the radiation and lack of oxygen on Mars. It will be a hard, lonely, and dangerous life, but they've all signed on for the greater good and to further scientific discoveries. The robots have been scouting for the presence of frozen water on the planet for years, and once the best location has been determined, it'll be mined and processed into drinkable water. At the same time, the construction robots are still hard at work building more landing and launch pads for future humans to arrive.

In 2042 the second crew arrives on Mars, this time containing around thirty astronauts. But a growing population means ever increasing needs. The priority of this mission will be to enhance the amount of food production on Mars. This mission will contain a hydroponic greenhouse, producing fresh fruits and vegetables from the locally-sourced water and supplementing the diet of pre-packaged astronaut foods. This will be essential to promoting the health of the astronauts and key to long-term survival on Mars.

In 2044 other countries are going to want to join NASA and SpaceX on the red planet. So instead of financing their own colonial missions they will likely decide to join the preexisting missions. This is around the time that allied space programs like those in Mexico, India, Europe, Canada, and Japan will realize that Mars Base Alpha is a success, and they'll cut their own deals to join. Soon, Starships will be coming from around the world, and Mars Base Alpha will start to look a lot more diverse. But now that humans are already on Mars, there's another challenge - getting them back to Earth for a return flight.

In 2046 there will likely be the very first return mission back to Earth. Using methane produced on Mars, the Starship rockets will take off and return to Earth. The base has now proven itself, and the entire world knows it can sustain life. Mars Base Alpha is now functioning as a full-fledged international research hub, with crews rotating in and out every two years.

In 2048 Mars is now producing its own air, water, fuel, and construction materials from locally sourced components. Huge 3D printers reduce the time it takes to create large construction components, and much of the hard work at the station is now fully automated with robots. The population of the base is approaching two hundred, and SpaceX continues to bring construction equipment for ongoing projects.

In 2050 Mars looks a lot different than it did when the project first began. It looks less like a red wasteland and more like a thriving outpost society. What started as a small organic garden now has grown into a big greenhouse, capable of supplying large amounts of fresh produce to the growing Martian population. But for the astronauts who want to eat meat, a fish farm will be established to provide some much-needed protein. Multiple species of freshwater fish, shrimp and other seafood will be raised in huge underground water tanks. But they're not going to be raising any chickens, goats, pigs or cows on Mars since they would consume too many resources and also risk contamination. A lab-grown cultured meat farm will be established to create small amounts of animal tissue for people who really want to eat a cheeseburger or steak.

Mars is thriving and the waitlist of people signing up to move there is longer than ever. The planet is starting to welcome wealthy tourists who want a taste of the ultimate adventure. The Mars base has been expanded with massive underground areas and tunnels. After testing the effects of space travel on children, NASA and SpaceX will decide whether to allow pregnant colonists to travel to Mars. As for kids born on Mars, that's a whole different question. The gravity and environment on Mars are very different from Earth. At only half the gravity, jumping on Mars sends people twice as high into the air. Basketball games on Mars are quite entertaining and golf is a much different experience with longer drives and greater distance between holes.

While adults with fully-formed bodies can well adjust to it, it's not as clear if the same would go for a child growing up on the red planet. Even if children could thrive on Mars, the trip back to Earth and adjusting to life with twice the gravity could be very hazardous for people's health. From Mars, It's a short trip to the two moons Phobos and Deimos. These moons of Mars haven't been deeply explored yet and may make for potential future sites for lunar bases. Until now, journeys beyond Mars were impossible due to the length of the voyage - but now it's possible to start on Earth, go to Mars, and journey from there. While the focus on Mars has been on developing it and researching its resources and history, there is one feature on it that likely attracts the attention of more visitors. It's called Olympus Mons, and it's the ultimate adventure for mountaineers. Standing taller than mount Everest, it's the tallest mountain ever found in the solar system. It's a long and treacherous hike, but that just means someone's going to want to be the very first person to conquer it. And for that lucky rich Earth tourist who conquers the mountain and lives to tell the tale, you can bet they'll be famous in more than one world.

In 2055 Mars is looking more and more like a human city with every passing day. The entire surface has now been explored and mapped out by humans. The area around Mars Base Alpha is now a developed spaceport capable of handling the traffic coming to and from Mars. Now, even bigger Starships have been sent. These larger ships are carrying hundreds of passengers at a time. They are not only bringing more humans to Mars, but heavier and more specialized building, scientific and life support equipment. Mars now has many more bases, mining depots, and research facilities. Between the official sites and the many tourist sites that are developing, people are going to need ways to get from one place to another. Mars will have a road system established, as well as a public transportation system. Elon Musk's interest in underground tunnels and train systems like the Hyperloop means that the mining droids have been busy digging a vast martian tunnel network. These are supplemented by buses and short-distance cars above ground.

In 2060 With the growing population on Mars, the planet is starting to look less like a small base and more like a thriving city. Many of the colonists on Mars work in factories on the planet, helping to construct things like solar panels and help perform the precise hands-on work that the robots can't. Life on Mars is becoming less of a sacrifice for the future of humanity and more of a desirable adventure for those able to make the long voyage. Many of the amenities on Earth are starting to become available on the red planet, and this added luxury is bringing more and more people all the time. New bases are being built and they soon start developing their own cooking styles, music, art and subcultures. It becomes increasingly important to protect the surface of Mars - with more and more of it being dotted with human life, a stray meteor strike from the asteroid belt could spell disaster. Underground tunnel systems are crucial to sustain life and protect from natural disasters.

In 2065 The population growth on Mars is increasing every month, and that means that making decisions for a large group is getting harder all the time. Up until now, many decisions were largely determined by SpaceX , NASA and governments back on Earth. But with a growing population of thousands, Martians will want to establish their own new form of governments. These governments are likely to resemble small local or city governments at first. They will most likely have a legislative council, an executive division, and a judicial system. They will direct criminal and civil disputes on the red planet. One of the next big ventures will be the process of terraforming Mars, which will be started by releasing greenhouse gasses in the Martian atmosphere and seeding the air. Others discuss the possibility of setting off nuclear bombs on the Norther and Southern poles of Mars, but this idea is met with a lot of fear and doubt. A massive stumbling block for Mars lies in the lack of any electromagnetic field.

Since Earth has a molten iron core, it naturally has a magnetic field that repels radiation from the sun. A lack of electromagnetic field would make any attempt at terraforming Mars vulnerable to solar radiation, so colonists are most likely going to stick to what works - large domed environments suitable for plant and human life and growth. New outposts will be established on the moons of Mars and spaceships will get faster. Mars will soon be covered with many massive domes containing cities of people.

Despite many concerns about the health and wellbeing of children, the first children will likely have been born on Mars by now - either planned, or more likely, through the natural result of humans sleeping with one another. But planned or not, this will be one of the biggest experiments yet on Mars. These children will be the first Martian natives, growing up in the red planet's weak gravity and thin atmosphere. It's possible that they'll live their whole lives on Mars, but if they plan on returning to Earth, that may become a difficult process - and they may have to do special exercises and physical therapy to adjust to Earth's increased gravity.

As the population of Mars continues to grow first from hundreds to thousands to the tens of thousands and beyond, the people there become increasingly self-sufficient. Many of the resident Martians may start questioning the need for control or leadership by Earth's countries. Some may even say it's time to take their fate into their own hands and declare independence from Earth. Like in the United States in 1976, Martians will join together to form a new declaration of independence saying:

"When in the Course of human events, it becomes necessary for one people to dissolve the political bands which have connected them with another, and to assume among the powers of Mars, the separate and equal station to which the Laws of Nature and of Nature's God entitle them, a decent respect to the opinions of all humankind requires that they should declare the causes which impel them to the separation. We hold these truths to be self-evident, that all people are created equal, that they are endowed by their Creator with certain unalienable Rights, that among these are Life, Liberty and the pursuit of Happiness. That to secure these rights, Governments are instituted among them, deriving their just powers from the consent of the governed, That whenever any Form of Government becomes destructive of these ends, it is the Right of the People to alter or to abolish it, and to institute new

Government, laying its foundation on such principles and organizing its powers in such form, as to them shall seem most likely to effect their Safety and Happiness, whether on Earth, on Mars or beyond."

Politics

Communities will require organized governance, so at some point, posts will open for decision-making for the colony, leading to political positions like governors, mayors, councilors, treasurers, trade authorities, space travel regulators, law enforcement, etc. This may be an organic process with informal and ceremonial gatherings and votes, but evolve into a more complex hierarchy as more people arrive on Mars.

How will Mars be governed, then? Strangely enough, the Starlink Internet Service Terms and Conditions (October

2020) states that the city would be self-governing. However, this would be in violation of the 1967 Outer Space Treaty (Article 8), which specifies "that the country that launches the rocket is accountable for all subsequent space activities". Does the Treaty take into account that even though Starship will launch from US territory, it will be launching citizens from around the world? Would they then be subject to American laws in space and on Mars? It may be a barrier too far for some international Mars travelers.

Furthermore, to plan for colonists to go out, own their own land, and establish their Marstead, that would also be impossible, as according to Article 2 of Outer Space Treaty, there can be no land ownership on Mars as no nation can appropriate a celestial body nor occupy a planet or part thereof and claim it as your own.

It would seem that the Treaty would require updating for the 21st Century. Plus, if nothing else, history has taught us that colonies do not stay so for long, and if freedoms are not forthcoming, then conflict will follow. So, the first job of any Martian official may be to renegotiate the Treaty so Mars has its own protecting its peoples and settlements.

However, in 2018, Neil deGrasse Tyson countered conflict arguments by stating that with planet-hopping technology, "a whole category of combat has the potential of disappearing totally. You don't have to wage a war in space; just go to another asteroid and acquire your resources." But you cannot move a whole surface-based city, and why should those working hard have to journey farther afield to mine resources or just live their lives so they can have some peace?

So, from the offset, Mars will have to have a contingency for its own constitution, informal rules, and eventual independence. Independent colonies around Mars will have their own rules. There must still be a cohesive order so those traveling between colonies and from Earth are welcomed. Starship must be a diplomatic ferry for peace, not war. It's the only way a multiplanetary civilization would last.

Communication

Starlink

With settlements on Mars potentially separated by thousands of kilometers, the best way to communicate will be by satellite. With SpaceX having launched their first Starlink satellite in 2019, the orbiting constellation of satellites is providing Earth locations with Internet signals sent through the vacuum of space, a mode of delivery much faster than fiber-optic cable. This offers the added advantage of reaching far more people and places. Starlink can enable email services, video calls, online gaming, streaming, and other high-data rate activities not usually available with other satellite Internet services. Marketed as the world's most advanced broadband Internet system,

Starlink also has a portability feature for its users.

Another benefit of Starlink is that it is a constellation of several satellites, rather than a single, geostationary satellite, and its closer orbit to Earth (at about 550 km) allows for a faster orbit time, thus reducing latency issues. It is ideal for rural and remote communities without access to connectivity services. And for Mars, such a system would be perfect.

As SpaceX president and COO, Gwynne Shotwell, explained, she expects Starlink broadband satellite constellation technology to be used for Mars as well. But Shotwell also stressed that Starlink will also have to "connect the two planets as well…" for "robust telecom between Mars and back on Earth." Such communication will help foster ties between the worlds, so Earth is not forgotten and Mars is not an isolated hope in the sky.

Deep Space Network

NASA's Deep Space Network (DSN) is the world's largest international array of giant radio antennas encompassing a sensitive, scientific, telecommunications system that supports interplanetary spacecraft missions. Operated by NASA's Jet Propulsion Laboratory (JPL), the DSN commands and oversees many of the agency's robotic space missions—like Perseverance on Mars—and also provides radar and radio astronomy observations.

The DSN has three facilities worldwide spaced equidistantly, approximately 120 degrees apart in longitude from each other, at Goldstone, near Barstow, California; near Madrid, Spain; and near Canberra, Australia. The chosen locations allow for constant communication with spacecraft as Earth rotates. So, as Starship departs Earth for Mars, communications can still be maintained and passengers can deliver and receive messages. Depending on the positions of Earth and Mars, radio signals can take anywhere between four to 20 minutes to reach each other.

ESA, Russia, and India also have their own DSNs, and with more countries accessing space or contributing to other space agencies, communication networks will grow.

Sun Orbital Relay Satellites

There will come a time when Earth and Mars are in conjunction with each other on the opposite side of the Sun. For a month, there will be no communication available until the Sun is out of the planets' way. But to avoid this communications blackout, satellites can be placed at strategic points orbiting the Sun to relay signals during this time.

Lasers

In 2004, ESA demonstrated the use of a high-beam laser as a communication method between its ground station in Tenerife, Canary Islands and its lunar orbiter, SMART-1. Such lasers, traveling at the speed of light, just like radio waves, can transmit 10 times or more data than radio waves.

Radio Mars

Colonists on Mars may go "old school" and use radio for local communication. Shortwave-radio use may depend on atmospheric conditions changing which frequency is used each day and allowing mostly voice rather than video communication. So, radios may become backup systems.

Work

There will be many jobs on Mars, even at the beginning for less-skilled people. Being a multitasker may be an asset on Mars rather than a specialist, but both skill sets will be required. The first Starship crews may have more specialized jobs, but the Martian frontier will be rough and ready for the right people to take a chance. The jobs on Mars will mirror what is on Earth—some more than others—but a few will be different. Here is just a taste of what jobs Starship's Mars immigrants can expect.

Perhaps the most important job after engineers would be an astrobiologist searching for evidence of life on Mars. As an interdisciplinary field, many scientists studying molecular biology, biophysics, biochemistry, chemistry, geology, and paleontology would be roaming around Mars looking for the telltale signs of life.

Roboticists would be required to build and maintain any robots on base for exploration, communication, construction, and manufacturing. While robots have been built and tested on Earth, Martian engineers may find they can build a robot decidedly more suited for Mars for exploring lava tubes, prospecting for water or minerals, and even keeping the base clean and tidy.

Mars will need those skilled at drilling wells for water, prospecting for water, or mining elements such as gold, silver, platinum, and more in much greater quantities than on Earth. These jobs may be tied to paying off work debts for some, if sponsored by mining companies.

Engineers of all types will be required, not just to help build bases, but to also drive ideas and innovations, expanding the first settlements and possibilities for survival. They would be the inventors, designers, and maintainers of machines, structures, and systems. Additive manufacturing (3D printing will be used on Mars, creating structures and products from plastic, metal, and ceramics) is a necessary service, so having the design and engineering skills for this would be ideal.

Construction crews would also be welcome, making bricks and cement and building anything, including shelters, boundary walls, water-well infrastructure, landing and transport shelters and pads, greenhouses, or even interior furniture. These workers would literally leave their mark on the landscape of Mars, perhaps even seen from space by the next passengers on Starship.

Drivers on Mars will have an important role as not everyone may be able to handle a large rover or be experienced with unpressurized vehicles. You wouldn't just be ferrying people around, but acting as a courier, explorer, roving reporter, or part of the search and rescue team. But drivers can also be the eyes and ears of the tourists, showing them the wonders of Mars.

If you have a knack for finding and restoring things, then salvage would be a worthwhile and profitable job. Crashed or expended rocket parts, lost vehicles, and reclaimed items could be recycled for your own use, returned to its owner for a fee, or scrapped for a fee.

Artists would also be needed for Mars, bringing interpretive meaning to their mission. Martian clay from the regolith can create ceramics just like pottery on Earth. Silicon dioxide makes up 40% by weight of Mars' regolith, and once the iron oxide is removed, can be made into glass. Industrial metals can also be made, like steel, silicon (for electronics), aluminum, and copper, through various processes. There would be plenty of material for artists to work from.

There may be military personnel (soldiers and scientists) on Mars, not specifically to protect colonists and settlements, but to experiment with new weapons, equipment and devices, combat techniques, and train for "space warfare". It may not be a popular choice and may skirt Outer Space Treaty Articles, but as on Earth, military innovations and systems have been known to enhance technological advancements, such as early versions of the Internet and GPS. Such military endeavors on Mars may help the colonists in their survival.

It goes without saying that medical professionals, law enforcement, academics/teachers of any field, accountants, insurance workers, and hospitality staff for hotels, bars, and restaurants would also be required. If you have a job, it will be useful on Mars.

Play

Some people say that 'news' stands for "News, Entertainment, Weather, and Sports". We humans live by these aspects on Earth, and so it will be on Mars. Not only will people be involved in reporting on these from locations on Mars, but also be involved *in* them.

There will always be newsworthy items to report on Mars, from new discoveries, to new Starship arrivals, new inventions and innovations, and of new water sources. Likewise, while Martian weather may not be as varied as on Earth, colonists would want to know when the dust storm season will begin so they can prepare their shelters and hunker down. As Mars' coldest temperatures are deadlier than Antarctica's, people would need to keep track of seasonal changes, or in more drastic cases, news of marsquakes, volcanic activity, and methane spurts will also want to be known by its audiences.

Sports can bring a new dimension in the Martian Olympics with new records set in field events. The 100-meter sprint may be a slow, undulating run, but the high jump and pole vault will produce stunning results. There may be crater races, sport volcano climbing, motor sport races akin to the Dakar rally, and the Great Valles Marineris Ultra Marathon. And of course, robot races and competitions will be an amusing challenge, as well.

In entertainment, Mars can produce their own TV shows and movies, with no CGI required for background shots so Martian movies will be genuine. Think of the costs saved in special effects, and the movies can be beamed back to Earth and the Moon for viewers. Documentaries about Martian natural habitats, mapping, and exploration programs viewed live, or features about people and their work and vision on Mars can be made, inspiring other Earthlings to make the next trip out to live or to be a tourist in space.

Economics

There will have to be some form of economic exchange system on Mars. People wouldn't work for free or just trade items. Currency is a system too ingrained into modern humans to dispense with straightaway. Many colonists may be short-term contractors working for five to 10 years before coming back to Earth. If they were employed or sponsored by a company, then they would want funds transferred to their accounts on Mars to use or even to protect from others on Earth. If they were getting funds on Mars, then they may want to transfer it to family on Earth. This may take the form of electronic transfer or most likely cryptocurrency for extra security.

Some form of taxation may also be levied on Mars for settlement upkeep and service provision. This would come from the Mars accounts, as well. But what of day-to-day life? Why would you need money on Mars?

Electronic transfers may still exist but may not be very secure in transmission. You would not want to hear that your hard-earned money sent from Earth was lost somewhere in the cosmos or a solar flare interfered with the transfer. You may also be the victim of discrimination if a bank on Earth didn't deal with Martian accounts or if currency rates or commission fees were robbing you of money. Your best bet would be to go full digital and away from centralized finances.

Cryptocurrencies are highly-encrypted, digital currency exchanged securely through a computer network that is decentralized, meaning it is not managed and controlled by a central authority like a government or bank. This may be the preferred currency-exchange format on Mars as peer-to-peer transactions may be more reliable and receiving funds from Earth would be secure. They would also be a form of investment or perhaps a Martian pension plan.

Apart from receiving a salary for working, there will be shop merchants, bars, restaurants, and other services which will require money. Crypto rates could be quickly established on Mars and used for daily transactions. It's a feature of Martian life Elon Musk has commented on, though others do not see the need for money on Mars. Some commentators have not foreseen that there will be things to buy, from necessities to luxuries. Not everything will be imported from Earth, but if it is, it will have to be bought. And cryptocurrency would cut through the "Which currency should we use?" issue, being able to be converted from one 'coin' to another.

As on Earth, there will be many types of cryptocurrencies on Mars, such as Musk's favorites, dogecoin or Marscoin, but many local coins would also be developed. This would enable Mars to have a thriving economy beyond Earth's control. There may still be a 'brick-and-mortar' Bank of Mars and limited coinage and bills for official reasons or for tourists, but Mars will be digital.

Trade may be conducted, but it would depend on what's for barter if a certain mineral is worth a rocket part. An engineer may trade a few hours of his work in a greenhouse for a week-load of fresh vegetables. And of course, wherever humans are, there are ill-gotten gains to be made, so a black market of goods will grow with contraband smuggled in on Starship or across Mars, selling secretly from unassuming outlets.

Mars may be a capitalistic society at first, but could develop a more broad-based economy to accommodate new arrivals and ideas. But without a strong, central Mars government, and with settlements wanting to be independent, having a cryptocurrency-based economy would engender trust and financial security.

Tourism vs One-Way Trips

You've paid your money (perhaps sold your house) to go to Mars. Have you decided to stay full-time on a one-way trip? Has a work contract to pay off loans been signed so you have to stay a minimum amount of time? Or are you wealthy enough to go for the thrill and return to Earth filled with a richness of the experience rather than material wealth? These will be decisions of a lifetime. Will you get married on Mars and raise a family? You will be starting an entirely new generation of humanity if you do. Imagine being the first to have a child on Mars, becoming forever enshrined in Martian history. It could be motive enough for some couples to try. Will there be a gender gap in colonists, jobs, marriage prospects, etc? Will Mars be ethnically diverse and protect any minority groups? What will be the age range of the colonists? Will retirees and mature people be as attracted to Mars as students and those in the more desirable 30-50 year age-group? Will there be an age limit? Perhaps if you're old enough to drink, vote, and marry, you would be eligible to travel to Mars. The cultural aspects of going to Mars have to be explored fully, understood by the outgoing colonists, and correctly implemented in laws and behaviors on Mars.

Tourists may pay less than a full colonist Starship ticket price, but a premium cost nonetheless to go to Mars for a limited amount of time. Colonists may be tour guides or act as hoteliers and hospitality staff, welcoming their Earthly cousins with gossip, news, and the best Martian tips for survival. As with Earth tourism, visitors will have to watch out for the tourist traps, with everyone wanting to go to Olympus Mons or the Valles Marineris. With the tourist season lasting 26 months before the next flight back, a lot can be achieved. If a tourist decides to stay on Mars, they would have to upgrade their ticket (so loopholes in tourist visas are not taken advantage of) and be in gainful employment so they pay their way on Mars.

Terraforming

Starship will bring humans to Mars, and humans will try to return Mars to its living, warm, wet early days by terraforming. Decades of study have revealed Mars was once a vibrant world until it lost its atmosphere and entered an ageless Martian glacial period. Mars' current average temperature is -55°C, but at the equator, it is between 0°C and 40°C.

Raising Mars' temperature by 10°C will create a feedback loop, releasing more absorbed CO_2 from the Martian polar caps and soil and create a thicker atmosphere within a century. Raising temperatures by 50°C would still be cold but survivable, with humans able to wear just breathing masks without a full space suit required outside. Higher temperatures would introduce rain and snow, (re)forming rivers, lakes, and falling water to Mars, again. The feedback process would, in turn, help to raise Mars' atmospheric pressure. Thus, the pressure difference between the interior and exterior of an inflatable, domed structure would be closer so domes could be built bigger and with no risks of collapsing.

Plants could be grown outside, soaking in the carbon dioxide and spread faster on the surface, producing more oxygen. With less CO2 will come lower temperatures, though more 'benign' greenhouse gasses could be introduced to maintain heat increases. And over centuries, Mars would be more Earth-like with no masks or suits needed for life support. Mars would be truly alive again.

Death on Mars

But just as on Earth, life on Mars will continue in the same manner with births, marriages, and deaths. Accidents will occur on base or in the Martian wilds. Human nature as it is, there may be a few suspicious fatalities, as we will be taking the best and worst of humanity's traits into space with us. Perhaps some will die of illness or old age. In worst case scenarios, a Starship may be lost. But this is the way of all explorations, of seeking the unknown. Not everything can be planned for, and there will be sacrifices along the way. But that would be no reason to quit or to criticize the endeavor.

Even Elon Musk, during a virtual conference he held on August 31, 2021, admitted there's a "good chance" the first settlers on Mars will die there. His meaning was more to the fact they would live out there lives there; not just crash in Starship, but live just as on Earth until death. But we would have to mitigate the notion of a life wasted on Mars and embrace the fact that the life chosen by a Mars colonist was their own, and they knew the risks. And if anything, their mortality on Mars will be elevated to immortality status in the history books as one of the first on the crimson globe.

But there could be another factor, illustrated by Musk himself:

"If I can go to Mars and be a human guinea pig, I'm willing to sort of donate my body to science. I feel like it's worth it for me personally, and it's kind of a selfish thing, but just to turn around and look and see Earth. That's a lifelong total dream."

If Musk feels this way, then so will many others. It may be a science thing or selfish motivation to die on Mars, to have lived and died on another world before returning to the stardust we are.

For a while, Mars and its immediate environs will be the furthest frontier of humankind. During the next 50 years of a million people settling on the Red Planet, other plans may be in motion for Starship and/or its successors. The solar system would be an open road. Where would we go next?

Chapter 8:

Beyond Mars

"People will visit Mars, they will settle Mars, and we should because it's cool."

-Jeff Bezos

The Moon and Mars won't be Starship's only solar system stops, if SpaceX and NASA have anything to say about it. If Starship proves successful, then interest will be ramped up for other destinations, like Venus and Jupiter. However, robotic spacecraft have taken years to get to their outer system destinations. Galileo took six years to get to Jupiter, but Magellan only 15 months to travel to Venus, so a shorter trip to the Evening Star may not be out of the question, with a flyby or maybe an insertion of a space station with regular supply reloads. But the short-term routes may be to Mars' moons for further study and then on to the asteroids for the main purpose of mining for resources.

Mars' System Exploration (Phobos and Deimos)

In mythology, Phobos and Deimos are the sons of Ares, the Greek god of war (identified with Mars—Rome's god of war). Phobos is the larger of Mars' moons, orbiting at an altitude of 5,989 km. It is oddly-shaped with one side being 27 km on its longest side. Its makeup is similar to C-type (or carbonaceous chondrite) asteroids. In fact, it could be a captured asteroid or debris from an ancient impact. It has no atmosphere, and its surface is covered in fine powder, the result of millions of years of micrometeorite impacts. Phobos is scarred with grooves, which scientists had thought were caused by an impact in its past. Now they believe from modeled data that the grooves are somehow caused by tidal forces acting against Mars. Phobos' orbit takes 7 hours and 39 minutes to revolve around Mars. In contrast, Deimos is the smaller moon at 15 km across at its longest side. It orbits every 30 hours at a much higher altitude than Phobos at 23,460 km.

NASA's Slow Approach

There are some astrophysicists who believe Mars' moons should be explored first before going to Mars proper. They want a slow, *slow* approach as landing on Mars would be complicated. They would use the moons as stepping-stones, like base camps, between the Earth and Mars so that equipment and resources could be sequestered in relative safety before descending to the Red Planet.

Their abundance of caution comes from the fact Mars is notorious for being a spacecraft wreck zone. Only nine of the 18 planned landers have actually landed safely. So, a 50% success rate for landing does not bode well for Starship, especially with it carrying a greater payload and 100 lives.

So, their idea is, instead of going straight from Earth to Mars, to land on Phobos and Deimos, set up stations, and ponder on how to get to Mars. It is not a very compelling argument from scientists who know Starship is in the works and has different capabilities and landing processes from remote spacecraft. Their agenda prioritizes visiting, exploring, and gathering samples from the moons before going to Mars. But this would add an unnecessary delay to Starship's mission; plus, how could 100 people safely stay on Starship or a base on the moons?

That also brings up another question: how did NASA come to this conclusion? In 2015, NASA JPL (Jet Propulsion Laboratory) engineers proposed "A Minimal

Architecture for Humans Missions to Mars", where spacecraft and personnel would land on the moon before landing on the planet. Compared to Earth, Mars may be smaller, but it still possesses a substantial gravity well. Launching from Mars into low orbit, a spaceship would require a velocity change of 3.6 km/s. And to return to Earth from Mars, a ship would require a velocity change of 6 km/s.

The four-stage proposal called for sending humans to Phobos to install infrastructure on the moon, then the astronauts would go down to Mars for a month (followed by a one-year mission) and then a more permanent stay. Figuring in Starship, the scientists reckoned it would require three launches to carry supplies, a moon habitat, and an Earth-return vehicle to Phobos, with the fourth rocket carrying the crew, who would be stationed on Phobos for 500 days, carrying out scientific works and perhaps visiting Deimos before returning to Earth.

From the scientific studies of Phobos and the shorter Mars landings, six more launches would be required to send more supplies to the planet, including a large, Mars-landing vehicle, which would stay in orbit. And then finally, a spaceship crew would journey to Phobos ready to land on Mars. Several crew members would then transfer to the descent vehicle, with a couple left on Phobos, while they landed on Mars for a month's stay. This plan is expected for the 2030s or 2040s with overlapping, crewed missions between Earth, Phobos, and Mars.

The plan illustrates NASA's slow forward thinking on Mars, as they are quite aware of Starship and its possible journey to Mars decades earlier. But the counterargument to this is it will still take Starship decades to build up that infrastructure for a million people on Mars, so why not practice with Phobos first? You can investigate landing areas, rescue procedures, and other science methods. Plus, it is a win-win for NASA, as they are SpaceX's largest customer, so they have hedged their bets with both their own slow mission and a cheaper, faster Mars mission with SpaceX.

Space Elevator?

But Starship may have a very practical reason for going to Phobos and that would be to investigate the feasibility of a space-elevator installation between Phobos and Mars. A 2003 NASA paper laid out the terms of such a possibility. With a 1,000-km tether lowered from Phobos, with its bottom end hanging over the planet, the tether would be moving through Mars' atmosphere at around half a kilometer a second, but more importantly, passing over specific points on Mars' surface twice a day. There could be such tethers from Phobos, and even a 3,000-km tether from Deimos, for moving payloads launched up from Mars' surface, and supplies sent down the elevator with hardly any propellant used. From Phobos, a one-way supply trip would take about two days. The energy for the moon-elevator system would be less than the launches from the surface, due to orbital mechanics.

However, while the idea of moon stations and space elevators sounds viable, the point of Starship is to deliver 100 passengers to Mars on each of the 1,000 Mars journeys. SpaceX might not go out of its way to forward a NASA mission to the moons. However, once Mars is settled, missions from Mars to Phobos and Deimos may be presented with elevators built to facilitate deliveries to other orbiting ships or spaceships docked at the moons on their way back to Earth, to the asteroids, or Jupiter. Why land on Mars to refuel when that can be done from the moons, with supplies and crew transferred over? Space elevators can be emergency exits from Mars if spaceships are not available or a secondary means for transferring goods. And as the technology matures on Mars, such technology can be transferred to Earth with space stations holding docked Starships, instead of moons. Starship could facilitate the opening of Mars' moons as stations and space-elevator bases and create the spaceport to the rest of the solar system.

Deimos' Tunnel Base

In 2013, Lockheed Martin Engineer, Josh Hopkins, recommended Deimos be the focus of the first missions to Mars. He argued that as Phobos and Deimos are tidally locked to Mars—displaying the same face to Mars—this would enable scientists on the moon base to be able to tele-operate rovers more quickly than their Earth counterparts. And sample return missions from Mars could be sent up a space elevator and transferred to a ship rather than to send a separate mission.

Deimos may provide a better prospect for such operations due to its higher altitude than Phobos, and it can observe 98% of the planet's surface. And because it travels less quickly across the sky, the length of the communication period with the surface is also superior to Phobos, claiming almost 60 continuous hours in comparison to 4.2 hours for Phobos. And in regards to launching to and from Deimos, it will require 400 m/per second less change in velocity.

Even more radical than space elevators would be to tunnel into Deimos, creating a so-called "O'Neill Cylinder" station. In his 1976 book, *The High Frontier: Human Colonies in Space*, American physicist Gerard K. O'Neill created the concept of a new space settlement which consisted of two counter-rotating cylinders. Both would be connected at each end by a rod via a bearing system, and the cylinders would counter-rotate in opposite directions to each other to cancel out any gyroscopic effects. Through this rotation, artificial gravity would be created.

In the case of Deimos, a tunnel would be drilled right through the center of the moon with the cylinders becoming a permanent habitat within Deimos. According to Jim Logan, the cofounder of the Space Enterprise Institute, who proposed the idea for Deimos, he suggested it would take 45,000 Starship launches to create an O'Neill

Cylinder with enough radiation shielding. But this idea was challenged by Alan Boyle from Geekwire, who calculated that Logan had underestimated the amount of radiation shielding that would be required by about a third; so, in fact, 150,000 Starship launches would be needed, instead. He even suggested Musk's previous endeavor, the Boring Company, could do the work. This is if Deimos is as porous as it is assumed to be. Also, water-ice reserves, precious metals, and minerals may also be discovered during the excavation. And for extra power, solar panels installed at Deimos' poles would receive constant sunlight.

There are exciting reasons to go to Phobos and Deimos, but settlement before Mars is not the goal of SpaceX. But once Mars is relatively settled, then colonists will begin looking at the moons for the next challenge.

Mining Asteroids

After the moons of Mars, prospectors, miners, and space explorers will start to eye the asteroids more seriously. Asteroids have been a source of fascination for decades due to the vast potential wealth each individual space rock stores within its body.

The types of asteroids scientists, engineers, and miners will be interested in fall into two main types: achondrites and chondrites.

Achondrites:

- Rich in platinum group metals (ruthenium, rhodium, palladium, osmium, iridium, and platinum).
- Located deep in the cores of planets.
- Remnants of early solar system metals coalesced into asteroids.

Chondrites:

- Rich in water for water, food, and fuel use by astronauts and asteroid miners.
- Help in creating an efficient radiation shield.
- Water can be transferred to Earth's moon base, Mars, and various stations.

The Main Belt

Between Mars and Jupiter lies the main asteroid belt with over one million asteroids at least one kilometer in diameter, with some larger than 10 km possessing masses up to two billion tons. Many of the asteroids will have an average of 200 million tons of iron, 30 million tons of nickel, 1.5 million tons of cobalt, and 7,500 tons of platinum. Besides the great economic wealth available, asteroids may also have huge stores of carbon, oxygen, hydrogen, water, and water vapor, like Ceres, Pallas, and Vesta, the largest of the asteroids.

From Mars, it may be relatively easy to travel to the asteroids, to start a mining industry and deliver a bounty of space riches back to Mars and Earth. Such a plan of action may relieve the harsh environmental impacts of mining on Earth, so affected countries and communities could start to reclaim lost lands due to mining. But it will not be easy to work on the asteroids, as scientists, engineers, and workers will have to carry their own survival supplies there. Will Starship be one of the first spacecrafts there? With its huge payload capacity, Starship may be able to transport supplies to various asteroids, such as life-support facilities, fuel production plants, and other vital resources.

For instance, the largest asteroid, Ceres, gave up its watery secrets in the form of hydrogen signatures to NASA's Dawn spacecraft in 2015 and the Hubble Telescope, which indicated Ceres could hold more water inside it than the whole of Earth. Smaller asteroids contain large amounts of water, too, but instead of investing in a large drilling infrastructure, techniques such as optical mining could harness the Sun's heat to bake the water out of the rock. But transporting water adds heavily to the payload, and thus, is very expensive to launch from Earth, costing on the order of \$9,000 up to \$43,000 to even send a water bottle into space—the reason why the ISS recycles all its water.

Akin to Mars' infrastructure of fuel propellant stations, the asteroids could provide this facility between low-Earth orbit and the asteroids. With the water split from the chondrite's hydrogen and oxygen sources, it can then be used for fuel processes shared between low-Earth orbit stations and the asteroids. And as one of the biggest costs in rocket launches is fuel, rockets leaving Earth can be lightly fueled and then refueled in space before going to Mars or the asteroids, decreasing the cost of rocket launches.

Near Earth Objects (NEOs)

There may be a way to test mining missions by traveling to the Near Earth Asteroids (NEO) first. There are over 27,000

NEO between Earth and Mars, two of the largest being 433 Eros at 17 km (11 miles in diameter) and 1036 Ganymed at 35 km (22 miles), but 90% of the NEOs are at least one kilometer in diameter and have low surface gravity. Adding to NEO numbers are known comets and meteoroids. It is these objects that cause a danger to Earth with near misses and/or collisions in the past. NASA's Galileo and Dawn robotic crafts have both obtained close-up imagery and new data from NEOs, while in 2010, Japan's Hayabusa became the first spacecraft to land on an asteroid and successfully return to Earth with samples. So, future mining robots and private commercial mining operations will be planned based on the data gleaned by these pioneering robots.

So, the race to claim the riches of asteroids is almost here. US Texas senator, Ted Cruz, believes that "the first trillionaire will be made in space." Peter Diamandis, the founder of the X-Prize competition for various tech development initiatives and innovations, created the asteroid company, Planetary Resources, in 2012, with Chris Lewicki and other partners in Washington. Soon after, US company, Deep Space Industries, was founded by Rick Tumlinson,

Stephen Cover, and partners. A few more contenders have also been established with varying development issues and progress. But one thing they have in common is they are not rocket-manufacturing companies. They are mining operators and will have to hire or buy rockets. And this is where Starship will come in. With its established reliability, low costs, and rocket power proven on Mars and its moons, and without viable competition in those areas, Starship will be in great demand.

NASA's Asteroid Dreams

NASA had entered the competition with high hopes. However, its prototype of the Asteroid Redirect Mission's (ARM) robotic capture module system was canceled in the 2018 NASA budget. The aim of this mission was to arrive at and return an asteroid into Earth's orbit to study and mine it in relative safety. And though the program was canceled, the NASA team has retained the key knowledge so mission plans are not lost.

But it may spell the end of NASA's mining mission to the asteroids.

Even so, investment banks and academic circles, like Goldman Sachs and Caltech, are ready to accept that asteroid mining will reap untold benefits for whoever invests in the space-mining industry and for who can overcome the technical challenges of getting to the asteroids and mining them. Caltech has estimated that an asteroid-mining mission could cost $2.6 billion; this compared to the billion-dollar set-up costs for a rare-earth-metal mine. However, while the set-up costs for a football-field-sized asteroid may be a billion dollars, that asteroid could contain $50 billion worth of platinum.

Before the canceled ARM, NASA had already launched OSIRIS-Rex in 2016, as a sample-return mission to asteroid 101955 Bennu. It reached Bennu in May 2021 and is due to return a sample to Earth in 2023. It could be an important first step in assessing the practical merits of asteroid mining.

The Asteroid Mining Aims and Pitfalls

But issues may remain in that rare metals are valuable because they *are* rare. If you start mining asteroids for all the platinum, how rare will platinum be after that without causing a price crash? Mining companies should also carry out operations in space and not return asteroids to Earth so as to avoid accidental crashes on or collisions with Earth. And who do you sell all the material to? But in asking these questions now, mining companies can adjust their operations to only mine and release precious metals when required, much like how oil is treated today. Asteroid reserves protocols may be required. These valuable resources may be traded for money or services or for creating a new economic paradigm on Mars.

In 2013, Martin Elvis, a Harvard University astrophysicist, calculated that perhaps 10 potentially metal-rich asteroids and 18 water-rich asteroids would be within Earth's technological reach at the time. But years later, with the advent of SpaceX, Elvis reassessed his opinion and increased that by a factor of 10. Starship is already altering the perception of reaching and mining the asteroids.

In all of this discussion regarding sending Starship to the asteroids, it is rather incongruous that Musk doesn't care for asteroid mining, calling the nascent industry "bogus" in 2003. Whether he holds the same conviction now is not known. But astrophysicist Amara Graps, the organizer of the biannual Asteroid Science Intersections with In-Space Mine Engineering Conference (ASIME) and founder of the Latvian-led Baltics in Space, believes Elon Musk will change his mind. Essentially, she believes that there's a disconnect between asteroid-mining companies and entrepreneurs/businesses. If businesses don't know about the benefits of asteroid mining, and if asteroid miners do not know how to advise on the science and financial rewards of asteroids to their potential best customers, then a gap will develop. The asteroid mining start-ups will require almost-instant cash flow to get started on the asteroids, so the sooner they and businesses can partner with each other in a mutually-beneficial relationship, the better.

If asteroid-mining companies are looking for inspiration for funding their start-ups, they can emulate Mitch Hunter-Scullion, CEO of the UK-based Asteroid Mining Corporation, who has taken to crowdfunding for his first asteroid-prospecting mission, called 'APS-1' ("Asteroid Prospecting Satellite 1"). Launch had been due in 2020—now due for 2025—from India due to cheaper costs. To further raise funds above their £2.6 million ($ 3.1 million) target, they will then sell the data they own. They have also developed several other missions for asteroid prospecting, exploration, and extraction.

But many think that asteroid mining gets a bad rap because of the hype and the distraction over the getting-rich aspects from the metal-rich asteroids. Other companies, like Planetary Resources, endorse this view, choosing first to target the water-rich asteroids and have the platinum-rich asteroids as secondary targets. With the water-rich asteroids, you have a confirmed revenue stream, as any rocket or customer would want water for fuel, and thus, you have a ready-made rocket-fuel station. It will take infrastructure to set up, but you will have guaranteed customers. Then, once you have the water market going, you can then concentrate on the metals and ensure a transport operation for deliveries.

There is a caveat to asteroid mining. As with any other space body, no one can legally own an asteroid. However, the US Space Act of 2015 does allow companies to own the products they mine. Luxembourg has passed similar laws, while the UK and Graps' Latvia are also eager to access the market. Though the wealth is made in space, it will be spent on Earth, more likely than not in countries which are friendly to asteroid mining.

Beyond the asteroids are the enticing gas giants, their moons, and the mysterious, icy realms on the outer edge of the solar system. Toward the Sun are Venus and

Mercury. As Starship opens up the inner solar system, it may also be able to reach the furthest limits, carrying humankind to new adventures.

Around the Solar System

While many people have focused on Elon Musk's mission to transport one million colonists on Mars, the rest of his 2016 Mars architecture presentation seems to have been forgotten. The overview may have been conceptual, but it is important in the understanding of Musk's long-term vision for both SpaceX and Starship.

The presented slides outlined missions to Saturn's moons, Enceladus and/or Titan, sampling the latter's methane and hydrocarbons available and also investigating Jupiter's moon of Europa, as well as its ocean, for signs of life, Kuiper belt objects, placing a fuel depot on Pluto, and perhaps visiting the Oort Cloud. In his remarks, Musk believes Starship can bring the solar system into humanity's grasp, establishing fuel depots around it, from Mars and on various moons. Starship and other rockets would be able to hop from space stations, to propellant depots, and to other moons and planets.

Starship could delve into the Kuiper Belt region with Pluto as a fuel depot, studying its formation and exploring for more. And the Oort Cloud, though much farther out, could be traveled to where the comets congregate, about 2,000 astronomical units (AU) from the Sun (whereas the Earth is one AU at 93 million miles, or 150 million kilometers). With water discovered throughout the solar system, fuel production would be the key to establishing bases and traveling to other worlds.

The advent of Starship is still yet to arrive, but its potential has been explored in far more detail and more publicly than any of NASA's outer-solar system endeavors over the past few decades. There is more belief than ever before that a mission to Mars is possible, but also the building blocks to the future of space travel by humanity. And yet, there are those who view Elon Musk, SpaceX, and Starship as false prophesizing, ego-aggrandizing, escapist hype.

Chapter 9:

Criticisms of SpaceX Starship Missions

"I would like to die on Mars. Just not on impact."

-Elon Musk

The hype over Starship irritates some people, experts, and laypeople, alike. Some don't like the fact Elon Musk has surpassed NASA's technological process within two decades. Technically, NASA is still ahead with its historic Apollo missions, but once Starship travels to the Moon, then SpaceX will be the preeminent space force. There are criticisms of Elon Musk for his incredibly overhyped mission timetables thrown out over Twitter without contextual justification. Additionally, while Starship has not flown, it has been hailed as the savior of humanity, a very premature action, according to other critics. But is it Musk's, SpaceX's, or Starship's fault they are lauded? Musk's and SpaceX's proven track record largely predicts Starship's success. But heavy lies the crown of the king, and there is always someone ready to topple him.

Starship Won't Work

It's hype. It's Elon Musk having another crazy idea. Starship doesn't have the ability to reach Mars. People should not go to Mars. These are some of the criticisms leveled at Elon Musk and SpaceX. They cannot fathom the logistics involved and think it is impossible to move 100 people in a spacecraft, let alone one million to Mars, which has not even flown yet. Some critics may be agnostic about Starship, but others are total nonbelievers, scoffing at the SpaceX faithful.

Even in this short time, they have forgotten Elon Musk's track record, not just with SpaceX, Tesla, and SolarCity, but also before that with Zip2 and PayPal. Musk has a winning streak everywhere he has been, disrupting industries for the better and generating new technological innovations. It seems very shortsighted and foolish to bet against Musk, and yet, people do. There may be some professional jealousy involved and dislike of his unfiltered social media posts and self-promotion, but in terms of raw-idea generation and production from those ideas, not many people can match Musk across different 'genres' of technology.

There is also a whiff of snobbery that an upstart rocket company had managed to beat NASA to the punch of creating reliable, fully-reusable rockets, having a coherent plan to go to Mars, and then having the cheek to corral NASA as a client, piggybacking them back to the Moon. There's a balking of the imagination in detractors when they sift through the mission specification of Starship's mission and decry the attempt to take one million people to Mars. The logistics will be too much, building a whole new industrial base and society on Mars will be impossible, and the obvious arguments that there will be no trees, no oxygen or nitrogen, no oil, and no real food. They expect people to be creatures of comfort forever on an increasingly uncomfortable Earth and forget there are those who would give anything to live on Mars as the ultimate statement of humanity. It will be hard, even more so than President Kennedy's proclamation in 1962. But it will be done.

The Wealthy Are Running Away From Earth

The worst criticism of the Starship missions is that the wealthy are running away, supposedly to shack up in luxurious space bunkers, leaving Earth to its own devices and demise. Prince William was particularly scathing after Jeff Bezos' Blue Origin flights as if Bezos was organizing the evacuation of Earth by the super rich. There's also criticism of Sir Richard Branson's Virgin Galactic and his superrich space tourist flights. Though the real venom is usually reserved for Musk and SpaceX.

The surface criticism is mostly over the cost of the tickets to space, which can range from $250,000 to $55 million. It seems the same critics forget that when passenger air flights started, tickets were exorbitant for years, well out of reach of the average person, but as competition increased and technology allowed for longer flights, the ticket prices and destination expansion, airplane travel became available for everyone. We are at the beginning of the space age phase where ticket prices are high and destinations low, but competition and technology are advancing at such a pace that within a generation, everyone will have the chance to have a suborbital flight or visit a space hotel or the Moon. Musk hopes that the cost of going to Mars will soon be the same as selling your house and moving to the Red Planet. It's that simple in economics.

But back to the "escaping wealthy" argument. Critics wield this statement as a truth, taking potshots at billionaires who are disrupting the space industry. This is coupled with the correlation that if they are working on space initiatives, they are not working for Earth. They forget that most of the billionaires are great donors to charity, whether announced publicly or not, and in Musk's case, his companies have been set up with the environment in mind (to varying degrees of success) with solar power, batteries for electric vehicles, saving resources with reusable rockets, and with giving humanity the option to expand the ever-growing population into space. For him, it is about humanity's continued survival, almost to the point of obsession since his brush with death from malaria in 2001.

The second part to this argument is, so what? Why can't the wealthy leave Earth? Their money will still be here, and in most cases, the wealthy aren't responsible for new innovations and exploration, so their expertise and experience can be taken elsewhere. This may be an expensive tourist gig for them, but they'll be living in airless, waterless, and cramped conditions with death lurking on the other side of the bulkhead. Their money can buy them a ticket to Mars, but their influence won't last beyond Earth's atmosphere. They will have to muck in with the rest of the colonists. Why begrudge them that? Detractors may think that the wealthy are abandoning Earth, but this is not an act of selfishness on their part, but one of *selflessness*.

The third part that critics miss or don't understand to articulate enough is we can do both in regards to saving the Earth and going to space. Opponents do not want to fundamentally know that there is no conflict in these notions whatsoever. It has never been about one or the other. It isn't just the wealthy who can save the world, so why use them as a scapegoat for social issues? We don't have to fix the Earth before going to space. But going to space *could* help fix the Earth.

And Prince William should remember his father's Prince's Trust encourages young people to challenge themselves, to find work, and to become leaders who may one day themselves or have children who want to go to Mars because of their experiences. Going to space will ultimately give them and their children unparalleled opportunities and the chance to work for a better Earth.

Even Musk, in 2018, stated that there's a 70% chance he would go to Mars, but that Mars is not an escape hatch for the rich. Musk's response to that idea was one's chances of dying on Mars was greater than on Earth and that the notion of coming back to Earth may be minimal. If critics believe most of the wealthy would fancy their chances on Mars, then they may have to think again. If the wealthy want to leave Earth, let them. It's their prerogative. If they stay on Earth, don't expect them to have exclusive abilities to save Earth. That's not their job. Whoever goes to Mars will just be accepting the mission to challenge themselves and to help humanity survive to its next best destiny.

Environmental Concerns

Elon Musk has consistently stated he wants his companies to be part of the sustainable solution to helping Earth. Having reusable spaceships manufactured with minimal waste would seem to go a long way in realizing this ambition. Critics, however, have noted a genuine potential for Starship to damage the natural and social environment around the launch sites.

As reported in Chapter 3, the Federal Aviation Administration (FAA) had cleared SpaceX for commencing Starship flight testing after consideration of the environmental impact of the subsequent testing and flights. SpaceX has 75 items of action to complete to lower the ecological impact of the surrounding environment by Starship launches. But it is Starship's exhaust that has caught the attention of environmentalists and critics. With Starship due to be tested frequently, it has the potential to create greenhouse gasses, with one scientist estimating that each Starship launch creates the same amount of greenhouse gasses as flying an airplane continuously for three years. While others have refuted the data due to rocket emission analysis issues, it may still be a significant carbon footprint. Be that as it may, prominent UK scientist, Martin Rees, called Starship's Mars mission proposal a "dangerous fantasy [and that] coping with climate change on Earth is a doddle compared to making Mars habitable." Greenhouse gasses are not the only issue with SpaceX propellant.

In June 2022, it was reported Falcon 9 fumes were toxic and "hazardous to humans", as claimed in a paper by Cyprus' University of Nicosia research team in the journal, *Physics of Fluids*. The RP-1 rocket fuel (kerosene-based with liquid oxygen), like SpaceX's Falcon 9, produces large amounts of carbon gasses and nitrogen oxides. It is a common fuel used by SpaceX, Blue Origin, and other commercial rocket manufacturers. But with SpaceX due to expand its launch capabilities, its carbon footprint could undermine their environmental credentials.

However, in both rocket exhaust cases, there are technological fixes, including having Starship refuel in space, leaving a minimal impact on Earth's atmosphere and on human health. New fuel types could be produced or Starbase operations moved to less populated areas. This is where innovation improves performance and new technology and/or processes. It also shows the transparency of the space industry and the tough safety regulations required to keep the rocket companies to a high environmental standard.

Social Disruption of Starbase

Not all Brownsville and nearby Boca Chica, Texas residents were happy to see SpaceX buy up land around them to establish Starbase operations. While the area would be enriched with new investments, jobs, and education, critics objected to the gentrification of the Brownsville area. Despite Elon Musk's announcements that SpaceX had chosen an empty area for rocket testing, their facilities and work would eventually incorporate Boca Chica into its city limits in Starbase. SpaceX bought out many of the Boca Chica residents who moved, despite other homeowners' resistance and doubts over the offers.

Initial plans by SpaceX revealed that only Falcon rockets were going to be tested and flown, but now Starbase is a Starship facility with more potential environmental impacts expected. Previous failed tests resulted in catastrophic explosions, scattering debris around Boca Chica Beach and disrupting adjacent wildlife reserves. So, the remaining residents are worried.

SpaceX's behavior may also encourage other tech or space companies to relocate to Boca Chica and surrounding areas as partners to SpaceX or to other small-town areas, interrupting many more residents' lives. And if SpaceX decided to open up new areas around the US for rocket operations, that may also fuel the critics' fire against the company. There will be a fine balance with locating rocket bases as the industry expands, and they will have to watch out for 'nimbyism' (Not in My Back Yard) protests. So, sensitivity to both environmental and social concerns will be paramount.

Humans Are Not Evolved for Mars

"We will never live on Mars, or anywhere else besides Earth." It is a grim statement by astrophysicist, Sylvia Ekström, and graphic designer, Javier Nombela, of the University of Geneva. They believe exploration of Mars should be the sole reserve of robots.

Their arguments include the fact that the human body has evolved over the millennia to survive Earth—its gravity, pressure, safety from solar and cosmic radiation, and the reasoning that when denied these Earth comforts, the body goes into "physiological stress". They are concerned over microgravity issues which would result in decalcification of bones, loss of muscle mass, weakening of the heart, fluid vascular system issues, increased thrombosis, and inner-ear disturbances. The researchers mention the exercise regimes astronauts can undertake to counteract some of these issues, but still believe Mars is an unattainable and undesirable place to be for humans. But they forgot one caveat to Mars missions—that humanity will have to evolve beyond Earth's prescribed limits in order to be a multiplanetary species. Human Martians will be different to humans on Earth, so will humans on the Moon, and those that may choose to be space-bound. Humans will still be humans, but may choose to evolve differently. That would be their choice and could take many generations to succeed, just like on Earth.

Their second argument that robots are better for Mars has been disproven by numerous tests on Earth and on the Red Planet itself. While robots can be programmed, it is rather limited in comparison to a human brain. In the BBC2 TV documentary, *Brian Cox: Seven Days on Mars*, Professor Cox visited NASA's Jet Propulsion Laboratory (JPL) to witness Perseverance's historic, five-kilometer journey in the Jezero Crater. The rover's aim was to navigate the crater, a previous lake—on Mars millions of years ago—toward its delta where life may have once existed. As per the title of the program, the rover had seven days to accomplish the task over rocky terrain. It would be both the fastest and longest a rover had ever traveled on Mars, and while it fell a bit short of its target, it had accomplished much. But, a human could have made the trip in a few hours or less. A human can detour to check out other sample areas or interesting targets while en route elsewhere. A robot does not have that luxury and will not until fully-autonomous versions are available. Yes, humans on Mars will cost more than robotic missions, having to have life support and supplies, but humans will always have better mobility, vision, and senses to explore Mars.

Where humans may be inferior to robots is with contamination issues. Robots have been sterilized to enter Martian habitats in their search for life. The Planetary Protection Act demands that contamination be avoided at all costs on non terrestrial bodies. Humans inherently carry their bacteria with them, which could interfere with the search for Martian life. Critics will insist that humans not enter suspected areas of special interest and to leave the exploration and testing to robots, which is a fair enough assessment. But it will not invalidate humans being on Mars.

Psychological Mars-fare

The psychology of space travel will also be a barrier to long-term ventures and colonization of Mars. French astronaut, Thomas Pesquet, has experienced the psychological pressure astronauts face on the ISS. It is enough when trained astronauts are worried about problems occurring in space, but that feeling will be intensified for those going to Mars. If there is a life-threatening issue on the ISS, astronauts can escape to Earth within three hours, but the Mars-bound ones could face long periods of extreme isolation for years. Pesquet knows that while academics and agencies try to recreate tough psychological situations on Earth (such as ESA's Mars 500 psychological isolation experiment), it will be hard to ingrain the real dangers of going to Mars while still on Earth. And multiply that by hundreds of untrained people going to the Red Planet, and there is a real risk of panic, fear, and hostility disrupting a Mars mission.

As mentioned before, one million people going to Mars will be a blip on the population of Earth, but they need to be totally dedicated to going to the Red Planet, not going on a whim or a dare. They must genuinely want to challenge themselves and change Mars and Earth for the better. We cannot have people changing their minds halfway through a journey to Mars, turning scared, and being disruptive, so the psychological tests will have to be strict and thorough. They should be trained to face their mortality, much like the Kobayashi Maru test in Star Trek; a no-win scenario to train cadets. Not everything will go wrong with Starship, but the passengers and crew will have to know they may not survive, but their sacrifice will not be in vain.

And that is the challenge of Mars: to have at least tried. We cannot be scared to change, to try and fail, and Elon Musk knows this. It is written into the DNA of his companies to try every iteration, with each failure shortening the path to success. Critics only see the failures and project doom from that, missing the little successes which are the real prize.

The Competition

The last critic comment goes to Elon Musk's bitter rival, Jeff Bezos, competing with Musk as the world's richest man and also against SpaceX with his Blue Origin.

Bezos has his critics, as well, but like Musk, he sees space as a way to expand humanity. However, his views and solutions to this are different. Bezos' plan is to develop huge, Earth-orbiting cities. Bezos would more likely endorse the Deimos plan to burrow into Mars' smaller moon and live there. He centers his space plan on Gerard K. O'Neill's 2019 study inquiring whether a planetary surface is the optimum location for human missions across the solar system. O'Neill's study concluded 'no', with three reasons:

1. Surface areas of planets aren't very extensive, and as populations grow, exploitable land surfaces may only be expandable by a limited amount.

2. The distance between Mars and Earth is far, over 200 times farther than Earth is from the Moon. So, with Earth and Mars aligning once every 22 months, the total voyage and stay on Mars will be years.

3. Communication between Earth and Mars will be difficult due to the distance. Signal lag could be up to 20 minutes.

With this in mind, Bezos purports to construct O'Neill-style communities in Earth's orbit. And to out-do Musk's one million people on Mars, Bezos claims space cities may be home to *one trillion people*. Now it was Musk's turn to be a critic, commenting on Bezos' plan:

"It doesn't make any sense. You'd have to transport massive amounts of mass from planets, moons, and asteroids in order to expand the colony. It'd be like attempting to construct the United States in the middle of the Atlantic Ocean!"

So, constructing cities in space may create their own environmental impact. But as with the stale dichotomy argument between Earth and Mars, there is no exclusive way to save humanity or the Earth. Moon bases, space stations, cities in space, and life on Mars can all be done. The more attempts, the better; the more multiple avenues for survival, the better.

There seems to be a constant either/or argument from critics, instead of exploring multiple strands of human expansion. Musk, Bezos, Branson, Diamandes, and the other commercial space companies are taking Darwinian principles into space. And only the fittest will survive. Rocket evolution will continue with competition driving humanity further toward the stars, leaving the critics behind to their own rash opinions.

Chapter 10:

Starship's Future

Whether from Elon Musk's announcements or from other quarters, the ship that has not flown yet has a great future projected for it. Starship has elicited such fervor and hope that even its early failures won't deter its inevitable rise into the cosmos. It's too beautiful to fail, too elegant to not grace the skies of Mars, and too young to not have a stellar future. But even before, during, and after any Mars mission, Starship will be drafted into other roles, establishing a legacy for decades to come.

The Near Future for Starship

Although named for outer space ambitions, Starship could also be applicable to Earth transport, enabling "superfast cargo transport" across the world, with trips from New York to Tokyo within an hour or across the Atlantic in a dozen minutes.

Starship will continue on from the Falcon stable of rockets, delivering the new generation of Starlink satellites around the world, when Falcon 9 and Falcon Heavy are phased out of crew flights, as well. A space analyst with the financial services company, Morgan Stanley, sees Starship and Starlink as a connected pair, with Starlink profits feeding back into Starship as it matures in improving its launch and payload capacities, reducing costs again. It's a virtuous circle, each nurturing the other in development. And whereas Falcon 9 flights can launch a maximum of 60 satellites, Starship can launch 400 into orbit. Starship will be part of the forefront movement to deliver a new kind of Internet service.

With this long-duration satellite launching capability and the reusability aspects, Starship will be well-tested before the mission to Mars. This will help SpaceX reduce costs in the shakedown missions. Musk believes he can get the costs down to under $10 million for an orbital launch, despite counterclaims by competitors, who cite Starship's multibillion-dollar costs and lower-than-expected demand from clients as a stumbling block.

Starship will also continue with private crew missions, such as the Polaris Program flights for billionaire Jared Isaacman, whose flights raise money for charities. He is undergoing astronaut training with SpaceX. Isaacman was the commander of the SpaceX's crewed Dragon Resilience flight, Inspiration4—their first private space flight in September 2021. And their partnership will grow into an effort to train Isaacman for Starship duty.

Other Starship tasks for the near future include:

- 2025—Starship HLS to demonstrate to NASA the human landing system for the Moon.
- 2026—Starship Crew (Heart of Gold) earliest potential crewed flight to Mars.
- 2029—Starship Cargo flight demonstrations for the earliest potential cargo flight to Mars (delayed from 2024).

The planned Starship tanker variant will also figure in the equation, refueling all the above Starship configurations in orbit, allowing for the same payload for flights to the Moon and Mars.

Part of the U.S. Space Force Fleet?

While SpaceX may not want Starship to be involved with the militarization of space, with the phasing out of Falcon 9 and Falcon Heavy (which have previously launched payloads for the US Space Force and the Department of Defense), Starship may be contracted for similar work.

Nothing has been announced, yet, but look to the US military to use Starship, less for satellite launches, and more as the ultimate troop or supplies transport, delivering them anywhere in the world within an hour. In January 2022, Space Force awarded SpaceX a $102-million, five-year contract to develop the Rocket Cargo program, which hopes to modify Starship rockets to carry around 90 tons (200,000 lb) of military cargo to anywhere in the world within an hour for tactical and humanitarian purposes.

Deep Space Starship

It's interesting to note, that at the moment, Musk does not see Starship as an interstellar vessel, despite its name. Such a voyage beyond the solar system would require greater velocities, so Musk favors an antimatter drive system. If Starship could be redesigned for this, retrofitted, or succeeded by a hypothetical Starship II, then Alpha Centauri could be within humanity's reach this century.

But instead of just launching humans, Starship may be capable of launching large space telescopes, such as the Habitable Exoplanet Imaging Mission, that can directly photograph planets outside the solar system. Or imagine a large telescope installed on Olympus Mons, 21 km up above Mars. Starship's power will enable scientists to see farther into the universe. And due to its capabilities, more and more planetary science researchers are beginning to imagine the possibilities with Starship, launching probes, equipment, and experiments quicker for cheaper costs, plus even achieving sample-return missions to the Moon and Mars.

Space Station & Tourism

Able to deliver crew and supplies to the ISS, and in the near future, to the Moon, Starship will become the go-to transport system for both professional and amateur commercial and industry astronauts. NASA will continue to use SpaceX for ISS duties, while other countries and agencies may follow suit. Enough billionaire "space tourists" or commercial astronauts have enjoyed their journeys on the ISS or with SpaceX, Virgin Galactic, and Blue Origin rockets. The next frontier will be open season for new tourists with expendable capital to enjoy the view of space.

If Yusaku Maezawa's dearMoon project is a success, then trips around the Moon will be a given. Space hotels have been a dream for many space entrepreneurs, and Starship's abilities will help kick-start the latent market. And with NASA's Artemis program, a triumphant return to the Moon's surface would also boost endeavors to build colonies there.

The ISS may not be the only space station around Earth for long. Starship could also find itself a platform for orbiting crews and visitors, letting scientists conduct experiments or even begin the trial stage for physical and mental tests for wannabe Martian voyagers. Long-term studies on the human body could be undertaken, not available anywhere else. It would be the center-point of Starship's Mars mission, with each of the selected 100 crew members having to live on board the Starship station for evaluation and their first taste of space life. It would instill in the Martian colonists a sense of home and stewardship over their Starship, as it will be their home away from home for almost a year until Mars.

Space Junk Collector

SpaceX's president, Gwynne Shotwell, has stated they are examining Starship as a potential space debris collector. Salvagers of human and robot teams could detect and recover space junk like "dead rocket bodies". Space junk that litters the area around Earth can be a great hazard to other satellites and spacecraft.

Cleaning up other people's space mess could be profitable should that nation want it back. It would be quite a feat as there are tens of thousands of pieces around Earth, but it would help space stations and future space cities survive. It will require precision flying to catch the pieces, but Starship's payload bay would be able to handle anything.

While Shotwell states Starship will collect other people's debris, it has been noted that SpaceX will end up contributing as much space junk as other agencies from its

Starlink constellation of satellites. With 60 Starlink satellites released each launch, the amount of debris could be huge, especially when Starship could launch 400 at a time. Shotwell is aware of the issues and explains they purposefully place the Starlink constellation in a lower orbit. Higher orbits encourage satellites to exist there for centuries and would be hard to repair if they failed, so the Starlink constellation was lowered as a whole so that errant satellites could be deorbited, decay, and burn up in Earth's atmosphere. Starship will have to be both a provider and cleaner solution.

Inspiration to Other Space Agencies

There is no doubt that SpaceX has inspired many people and those in the space industry, alike. While there is a competitive spirit amongst the current commercial space industries, many more are on the rise using such technologies like 3D printing to craft their rockets and payloads for greater efficiency, payload specs, and cost reduction.

Companies like Relativity Space are using robotics, software, and patented 3D-printing technologies to vertically integrate and digitize their manufacturing processes at their Stargate factory. Their aim is to 3D print and launch a rocket on Mars. Others, like Orion Additive Manufacturing, are creating 3D printers for use in space, working with ESA on projects to develop high-performance polymers for CubeSat production.

With start-ups like these, the commercial space industry could totally revolutionize the space experience. Interlinked technologies and shared innovative processes could lead to a viable Mars colony with the tools and belief that the colony will survive.

The Next Million People to Mars

It may take two generations to complete, but Starship could deliver one million people to Mars. It would not stop there, though. Whether Earth continues to suffer the worst of climate change, social upheavals, or financial disparities, there will always be those who want more, to challenge themselves and start anew.

The second wave of colonists may join the first wave or strike out on their own. The original colonists would by then have figured out an optimal way to live and work, with some having children. The terraforming process may have begun, and with more people, it may be accelerated. New languages and cultures may develop, and political and environmental groups may form wanting to protect the Martian way of life.

This will still be a young population with a lot to learn about their new world.

Mars will be shaped by human hands, but also in turn, transform the humans into Martians. And if Mars' original life-forms are found, would cohabitation be in the cards if they are found alive, or will Mars have a dead past? Humanity could find itself not just a multiplanetary species, but also a cohost to a planet, which would be a crowning achievement for Starship.

Conclusion

Starship is a revolutionary ship about to carry out an audacious mission. Starship could become the first spacecraft in history to land on three different worlds: Earth, the Moon, and Mars. One million people on Mars will be an unprecedented achievement for Elon Musk and SpaceX, and when the dust has settled, we will be a multiplanetary species.

In looking at the specifications of Starship, we can see the iterative development of the craft. Its skin is the purposeful silver of stainless steel, a perfect match for its outer space adventures. A belly of black for heat shielding lets it coast, then flip to vertical for landing. Its monstrous Raptor engines provide unmatched rocket power. And with the payload bay, a cavernous maw for the transporting of humans and equipment for the voyage to Mars, Starship was made for this.

Flight testing will follow soon after the FAA's announcement, and the flight no doubt will be streamed online and followed by millions, with the silver arrow punching through the air. And soon, the Moon will follow, bearing both commercial missions and NASA around and to the surface, respectively. The new frontier will have opened, and investment in space will grow as people realize the potential before them.

We know the 'where' and have seen the 'what', 'how', 'when', and 'why' we need to go to Mars. It's for humanity's survival. The one million colonists will need not just the supplies, life support, shelter, and protection, but also from Earth, our faith, patience, praise, thanks, admiration, and hope. They will be sacrificing everything, even some of their lives, to ensure humankind has an escape hatch, not just for the wealthy, but for everyone.

And beyond Mars are the moons, asteroids, and space stations enabling further living and working deeper in our solar system. It will be the stuff of science fiction, but in this case, science *fact* will be the reality. And for the critics—for there will always be critics—they will only spur on the colonists and rocketeers. Earth and Mars would have more to worry about and to celebrate as humanity learns to live among the stars.

So, what is the future for Starship? It will be a long and emotional journey with humans traveling to places never thought possible.

Long after Elon Musk has gone, whether on Mars or on Earth, Starship will be remembered as the founder's dream to do something different, to demand a new way of space travel, and to create a new, multiplanetary species in the galaxy.

Even before it has finished, Starship has become a legend.

References

Al-Sibai, N. (2022, May 20). *SpaceX
rocket fumes "hazardous to humans"
and the climate, scientists warn.*
Futurism. https://futurism.com/the-
byte/spacex-rocket-fumes-toxic

Arevalo, E. (2021, June 14). *SpaceX
plans to hire a "Space Operations
Training Engineer" to establish an
astronaut training program.*
TESMANIAN.
https://www.tesmanian.com/blogs/

tesmanian-blog/spacex-astronaut-
training

Asteroid Mining Corporation.
(2022). *Missions*. Asteroid Mining
Corporation.
https://asteroidminingcorporation.
co.uk/missions

Bergan, B. (2022a, June 13). *We're
going to Mars! SpaceX wins FAA
approval for Starship launch.*
https://interestingengineering.com
/mars-spacex-faa-starship-launch

Bergan, B. (2022b, June 14). *Elon Musk says Starship will be "ready to fly" into Earth orbit next month.* https://interestingengineering.com /elon-musk-starship-spacex-orbit

Brian Cox: Seven Days on Mars. (2022, June 17). BBC2.

Cain, F. (2019, September 18). *Want To Explore Mars? Send Humans To The Moons Of Mars First: Phobos And Deimos.* Universe Today. https://www.universetoday.com/14 3438/want-to-explore-mars-send-humans-to-the-moons-of-mars-

first-phobos-and-deimos/

Crossman, F., & Zubrin, R. (2005).
*On to Mars. vol. 2 Exploring and
settling a new world.* Apogee.

Delbert, C. (2020, December 29).
*Elon Musk says Mars settlers will
use cryptocurrency, like
"Marscoin."* Popular Mechanics.
https://www.popularmechanics.co
m/space/moon-
mars/a35085273/elon-musk-says-

mars-settlers-will-use-
cryptocurrency-like-marscoin/

Delbert, C. (2021, February 18). *Elon
Musk says settlers will likely die on
Mars. He's right.* Popular
Mechanics.
https://www.popularmechanics.co
m/space/moon-
mars/a33900282/elon-musk-says-
settlers-will-die-on-mars/

Glester, A. (2018, June 11). *The
asteroid trillionaires – Physics
World.* Physics World.

https://physicsworld.com/a/the-asteroid-trillionaires/

Hartmann, W. K. (2003). *A Traveler's guide to Mars : the mysterious landscapes of the red planet*. Workman Pub.

Horn, J. (2015, May 15). *Shackleton's Ad – Men wanted for hazardous journey*. Discerning History. http://discerninghistory.com/2013/05/shackletons-ad-men-wanted-for-hazerdous-journey/

Houser, K. (2020, January 2). *Elon Musk hints that a Cybertruck is headed to Mars*. Futurism. https://futurism.com/the-byte/elon-musk-hints-cybertruck-mars

Mars, K. (2016, August 17). *Gateway*. NASA. https://www.nasa.gov/gateway

Montag, A. (2018, March 16). *Elon Musk explains his motivation to succeed: "There need to be things that inspire you."* CNBC; CNBC.

https://www.cnbc.com/2018/03/16/
elon-musk-on-inspiration-and-
success.html

Murphy, G. (2010). *Mars: a survival
guide*. Abc Books.

Orwig, J. (2015, April 21). *5
undeniable reasons humans should
go to Mars - Business Insider*.
Business Insider; Business Insider.
https://www.businessinsider.com/5
-undeniable-reasons-why-humans-
should-go-to-mars-2015-
4?r=US&IR=T

Roach, M. (2011). *Packing for Mars :
the curious science of life in the
void*. W.W. Norton.

SpaceX. (2021). *Starship*. SpaceX.
https://www.spacex.com/vehicles/s
tarship/

Starlink. (n.d.). Starlink.
https://www.starlink.com/satellites

Tangermann, V. (2020a, October
22). *SpaceX president: Starship
could help pick up space junk*.

Futurism.
https://futurism.com/spacex-
president-starship-space-junk

Tangermann, V. (2020b, October
22). *SpaceX will build Starlink-like
constellation around Mars, its
president says*. Futurism.
https://futurism.com/the-
byte/spacex-starlink-like-
constellation-mars

The Mars Blueprint. (2022, June 26).
*For the love of Robots… A Mars
base needs Tesla's Optimus*. The

Mars Blueprint.

https://themarsblueprint.com/to-
make-a-mars-base-possible-spacex-
needs-the-tesla-optimus-
robot/?fbclid=IwAR1y8nNj9UNVN
gr68ueYZqmNariUZJbwUZsO-
4m_XUjPtA32gYjRW4_jgho

Tzinis, I. (2020, March 30). *Deep
Space Network (DSN)*. NASA.
https://www.nasa.gov/directorates/
heo/scan/services/networks/deep_
space_network

Vance, A. (2015). *Elon Musk : How the billionaire CEO of SpaceX and Tesla is shaping our future.* Virgin Books.

Wall, M. (2016, September 28). *SpaceX's Mars spaceship could explore the entire solar system, Elon Musk Says.* Space.com. https://www.space.com/34219-spacex-mars-spaceship-solar-system-exploration.html

Wikipedia Contributors. (2019a, October 15). *SpaceX Starship.*

Wikipedia; Wikimedia Foundation.
https://en.wikipedia.org/wiki/Spac
eX_Starship

Wikipedia Contributors. (2019b,
October 25). *We choose to go to the
Moon*. Wikipedia; Wikimedia
Foundation.
https://en.wikipedia.org/wiki/We_
choose_to_go_to_the_Moon

Wikipedia Contributors. (2021a, May
27). *SpaceX Mars program*.
Wikipedia.

https://en.wikipedia.org/wiki/Spac
eX_Mars_program

Wikipedia Contributors. (2021b,
June 8). *SpaceX Raptor*. Wikipedia.
https://en.wikipedia.org/wiki/Spac
eX_Raptor

Wikipedia Contributors. (2021c,
June 23). *Starship HLS*. Wikipedia.
https://en.wikipedia.org/wiki/Stars
hip_HLS

Wikipedia Contributors. (2021d,
October 5). *Space Exploration*

Initiative. Wikipedia. https://en.wikipedia.org/wiki/Space_Exploration_Initiative

Wikipedia Contributors. (2022a, January 3). *Mars Direct*. Wikipedia. https://en.wikipedia.org/wiki/Mars_Direct

Wikipedia Contributors. (2022b, May 28). *List of SpaceX Starship flight tests*. Wikipedia. https://en.wikipedia.org/wiki/List_of_SpaceX_Starship_flight_tests

Zafar, S. (2022, June 13). *Welcome To Mars City: Complete map of this Elon Musk's ambitious project - The Globe's Talk*. The Globe's Talk. https://theglobestalk.com/welcome-to-mars-city-complete-map-elon-musks-ambitious-project/?fbclid=IwAR3ooB768b7TYKB2EXPHWIMe-bEG-_R8o9FOeHrw2FKzjiYYpPaz7a3DfJk

Zubrin, R. (1996). *Case For Mars : the plan to settle the red planet and why we must*. Free Press.

Zubrin, R. (2004). *Mars on Earth :
the adventures of space pioneers in
the high Arctic*. Jeremy P.
Tarcher/Penguin.

Zubrin, R. (2008). *How to live on
Mars : a trusty guidebook to
surviving and thriving on the Red
Planet*. Three Rivers Press.

Zubrin, R. (2019). *The case for space
: how the revolution in spaceflight
opens up a future of limitless
possibility*. Prometheus Books.

Zubrin, R., Crossman, F., & Mars Society. (2002). *On to Mars : colonizing a new world*. Apogee Books.

www.ingramcontent.com/pod-product-compliance
Lightning Source LLC
Chambersburg PA
CBHW071732150726
47998CB00005B/1610